Extraordinary Engineers

Volume 2

By Dr. J. A. Sanchez
Extraordinary Engineers

Extraordinary Engineers:
Extraordinary Engineers Volume 2
Female Engineers of This Day and Time
Radiant Pearls Publishing
2024

All journeys in this book are contributions of the extraordinary engineers.
Printed in the U.S.A.
ISBN: 979-8-9877073-0-2 (paperback), 979-8-9877073-1-9 (hardcover), 979-8-9877073-2-6 (eBook), 979-8-9877073-3-3 (color paperback), 979-8-9877073-4-0 (color hardcover)

Cover art done by Get Covers https://getcovers.com/
Editing done by Jeanette Crystal Bradley.

Contents

Foreword

It is with immense pride and gratitude that I introduce this remarkable book celebrating the extraordinary contributions of women in STEM (Science, Technology, Engineering, and Math). Having the privilege of working alongside the author, who is not only a dear colleague, but also a cherished friend, I can attest to her unwavering commitment to excellence and her passion for continuous learning.

We met 19 years ago when Justina joined TÜV SÜD America. We have worked together in various ways, but the most notable was our work together within our company's Women's Network. As officers of this invaluable organization, we shared a vision of empowerment, advocacy, and support for women in a male-dominated field – a vision that continues to drive us forward to this day. Through our collaboration in CHAMP, a high achievement employment development program rooted in the principles of continuous improvement and learning, we honed our skills, expanded our horizons, and forged lifelong bonds that have enriched our professional and personal lives immeasurably.

Through our shared experiences, one theme has remained constant: our unwavering belief in the power of mentorship to inspire, guide, and elevate individuals to reach their full potential. As mentors and mentees, we have both been beneficiaries of this transformative relationship, sharing insights, and championing each other's successes every step of the way. It is this spirit of mentorship and collaboration that infuses the pages of this book, illuminating the diverse experiences, perspectives, and achievements of women from around the world.

From the young girl who dreams of becoming an astronaut to the seasoned researcher breaking new ground in artificial intelligence, every woman in STEM has a story worth telling. In these pages, you will encounter stories of resilience, innovation, and triumph – testaments to the spirit of women who have dared to dream, to defy expectations, and to shatter barriers in pursuit of their passions. From groundbreaking research in biotechnology to revolutionary advancements in artificial intelligence, the women featured in this book embody the essence of courage, curiosity, and perseverance, inspiring us all to reach for the stars.

As I reflect on our shared journey and the countless lives we have touched along the way, I am filled with gratitude for the opportunity to stand alongside the author in championing the cause of women in STEM. To Justina, I extend my heartfelt congratulations and deepest admiration. Your passion, your dedication, and your unwavering commitment to empowering women in STEM serve as a beacon of hope and inspiration for us all. May stories of the co-authors of this book ignite a flame of curiosity, ignite a spark of inspiration, and ignite a passion for discovery in the hearts and minds of readers around the world.

So let us celebrate the women in STEM – past, present, and future – and let us recommit ourselves to building a world where every woman has the opportunity to thrive, to succeed, and to make her mark on the world.

With deepest respect and admiration,
Theresa Glenna
GMA Senior Technical Sales Manager

Extraordinary Engineers
Volume 2

Extraordinary Engineers

An Engineer is a Princess Who Can Build Her Own Castle

Coauthored by Kelsey Kirrene

Background

Growing up in rural New Jersey, I was the quintessential girly-girl with a love for all things pink and sparkly. But beneath the surface of Barbies and glitter, a budding engineer was already beginning to show. This duality defined my childhood, as I seamlessly transitioned between playing with dolls and building intricate structures with toys like Knex and Legos. Building rocket ships and roller coasters for my stuffed animals became the norm, foreshadowing a future where creativity and technical prowess would coexist, and my passion for building things to help others could one day be realized.

In high school, I had the privilege of attending the Governor's School for Engineering & Technology, which was my first experience working with engineers and other students who were passionate about science, technology, engineering, and mathematics. This experience became the gateway to a world where my fascination with innovation and problem solving took root, through my high-altitude balloon photography project that failed - but more on that later. Little did I know that this was just the beginning of a journey that would take me from the halls of academia to the forefront of cutting-edge research.

Carnegie Mellon University became my academic haven, where I not only earned my bachelor's degree in mechanical engineering, but also delved into the intricacies of global engineering. This minor reflected my belief in the broader impact of technology on a global scale. And because life is all about balance, I added a touch of cultural finesse with a certificate in Spanish language and culture.

During my undergraduate years, my love for the underwater world led me to dive into research on mobility platforms for underwater robotics. The thrill of exploring uncharted technological waters fueled my passion for cutting-edge advancements. Simultaneously, my commitment to sustainability manifested as I worked on designing modular and maintainable solutions to enhance the fuel efficiency of large ships. This is a testament to my dedication to making a positive impact on the environment.

An internship in Japan added an international chapter to my story, providing invaluable insights into different engineering landscapes. The experience broadened my horizons and deepened my understanding of the interconnected nature of global innovation.

Post-graduation, the aerospace and defense industry beckoned, and I eagerly embraced the challenges it presented. From outer space satellites to cybersecurity on airplanes, I navigated through diverse technological domains, showcasing my versatility and commitment to staying at the forefront of innovation.

As a doctoral candidate, my current chapter involves delving into the nuanced world of ethical leadership styles within the U.S. based technology workforce. This academic pursuit mirrors my belief that true progress extends beyond technical excellence, touching upon the ethical dimensions of leadership in the ever-evolving realm of technology.

My journey is a tapestry woven with threads of passion, innovation, and a deep-seated commitment to making a positive impact in the world of engineering and technology. As I continue to explore new horizons, push the boundaries of knowledge, and inspire others, my story stands as a testament to the harmonious coexistence of creativity and technical brilliance.

Why Mechanical Engineering?

Mechanical engineering, for me, was not just a career path; it was a natural convergence of my love for physics, hands-on learning, and the sheer joy of crafting. It became the canvas where my passion for DIY and the intricacies of engineering seamlessly blended, creating a vibrant and fulfilling journey.

From a young age, the allure of understanding the mechanics of the world around me fascinated me. Physics became a playground of curiosity, where the laws governing motion and energy sparked a flame of fascination. This intrinsic interest laid the foundation for a future in mechanical engineering—a field where the principles of physics come to life through the creation of tangible, functional marvels.

What set mechanical engineering apart for me was the hands-on aspect. I was never content with just theories and equations on paper;

I craved the tangible, the experiential. The prospect of rolling up my sleeves and getting down to the nitty-gritty of building, crafting, and experimenting drew me towards this dynamic discipline.

But it wasn't just the technical side that captivated me. I've always had a soft spot for arts and crafts, and mechanical engineering turned out to be the perfect canvas to blend my love for creativity with the precision of engineering. The marriage of form and function became a guiding principle as I navigated through the intricacies of designing and creating.

College life introduced me to the hallowed halls of the maker space—a haven for dreamers, tinkerers, and creators. Here, surrounded by an array of tools and materials, I found my sanctuary. The maker space became my playground, a place where my ideas could evolve from concepts to tangible prototypes.

I spent countless hours immersed in little crafting projects, each one a testament to the marriage of my artistic inclinations with the technical prowess of mechanical engineering. Whether it was designing a more efficient gear system or creating a whimsical kinetic sculpture, every project became a canvas for self-expression and innovation.

The maker space wasn't just a place for solitary endeavors; it became a community. The exchange of ideas, the collaborative spirit, and the collective buzz of creativity fueled my passion even further. Soon, I found myself not only a frequent visitor but a Teaching Assistant, entrusted with the responsibility of supervising the space. This role allowed me to not only share my knowledge, but also witness the spark of inspiration in others—a deeply gratifying experience.

What resonated most with me in mechanical engineering was the tangible impact it had on the world. Every project, every creation had the potential to improve lives, solve problems, or simply bring joy. The prospect of being a creator, an innovator, and a problem solver became my driving force.

Mechanical engineering, for me, was not just a profession; it became a way of life. It provided a platform where my love for physics, hands-on learning, and arts and crafts could harmoniously coexist. The maker space, with its whirring machines and the symphony of creativity, became the backdrop of my college journey—a journey that allowed me to not only pursue my passions but also inspire others to embark on their own path of creation and discovery.

From the Lab to the Field

Navigating the transition from the controlled environment of the lab to the dynamic challenges of fieldwork has been a testament to the evolution of my curiosity and honed skills. The laboratory served as my training ground, where I meticulously crafted and fine-tuned robots, delving into the intricacies of their design and functionality. Armed with the knowledge and expertise gained in this controlled setting, the allure of real-world applications beckoned. Stepping into the field, I brought with me not just the tools of the trade but a curiosity that thrived on the unpredictable and the unexplored. The transition became a journey of adapting theoretical prowess to the practical demands of the field, where my skills became a bridge between the meticulously planned experiments of the lab and the unpredictable terrain of the outside world.

Venturing into the field, particularly in the rural port towns of Japan, brought a unique blend of challenges and rewards to my work in building and testing robots. Clad in safety equipment that transformed me into a high-tech explorer, I embraced the real-world environment with a determination to push the boundaries of innovation. Gearing up for tests meant more than just strapping on a helmet; it was a ritual of donning gear that would protect me and enable me to navigate the uncharted waters—sometimes quite literally.

The rural port setting presented an arena where the theoretical met the tangible, where the robots I crafted in the lab would encounter the unpredictable realities of the maritime world. The tasks involved not just running tests but immersing myself in the elements, with safety precautions translating into more than just a checklist. Waterproof boots became my steadfast companions, often getting a bit more acquainted with the water than planned, as I navigated through the port's uneven terrain.

The beauty of this hands-on experience lay in the authenticity it brought to the testing process. No simulation could replicate the nuances of a real-world maritime environment, and it was in these field excursions that the true capabilities and limitations of robots came to light. Whether it was assessing their mobility on different surfaces or evaluating their performance in varying weather conditions, the data collected in these moments was invaluable.

There's a certain thrill that comes with fieldwork—a palpable connection to the real-world implications of the work we do. It's not just about the numbers on a screen; it's about seeing the robots navigate, adapt, and interact with the surroundings they were designed to navigate. The challenges, from the unpredictable weather to the ever-changing landscape of the port, became integral parts of the testing process.

While the boots occasionally got wet, it was a small price to pay for the insights gained in the process. Those moments of navigating the port, clad in safety gear, were not just about testing robots; they were about pushing the boundaries of what was thought possible, refining designs, and uncovering innovative solutions to real-world challenges.

Working in the field became a journey of discovery, where every wet boot and safety precaution symbolized a commitment to bridging the gap between theoretical concepts and the practical demands of the maritime world. It was a reminder that innovation doesn't unfold in the sterile environment of a lab alone; it thrives where the challenges are real, where the boots may get a bit wet, and where the true potential of robotics is revealed in the unscripted dance with the elements.

Why Aerospace and Defense?

Embarking on a career in the aerospace and defense industry has been more than a professional choice for me; it's been a deeply personal and meaningful journey that resonates with the echoes of service ingrained in my family history. Growing up in a military household, where service was not just a duty but a way of life, I inherited a passion for contributing to the greater good. My mother's engineering endeavors during Desert Storm, my grandfather's service as an engineer in the Air Force during World War II and the Korean War, and now my brother's role as a helicopter pilot in the U.S. Army, all forged a path of service that became an intrinsic part of my identity.

Choosing a career in the aerospace and defense industry became a conduit for me to extend that legacy of service. It's not just about technology and innovation; it's about supporting those on the front lines, the warfighters who sacrifice so much to keep us safe. Building a bridge between cutting-edge technology and the protection of those who serve, I find profound meaning in contributing to advancements that enhance their safety and effectiveness.

Currently immersed in the realm of remote sensing within the space industry, I find myself at the forefront of developing payloads that utilize an array of sensors to capture the complexities of both Earth and space. This technology, with its myriad of civil and military applications, extends its impact far beyond the confines of the industry. From weather forecasting to missile detection, the data generated from these payloads plays a crucial role in ensuring the safety of individuals every day.

The connection between my work and its real-world impact is tangible and immediate. It's not just about the allure of space exploration; it's about the tangible difference the technology makes in the lives of people, whether civilian or military. Being a part of a field that contributes to weather forecasting, aiding disaster response efforts, and enhancing national security is not just a career; it's a vocation with a purpose.

In the aerospace and defense industry, I've found a way to blend my passion for service with a commitment to technological advancement. It's a space where innovation meets duty, where the work I do resonates with a familial history of service, and where every project becomes a tribute to those who dedicate their lives to keeping us safe. This career isn't just about building payloads or pushing the boundaries of space exploration; it's about standing in solidarity with the warfighters, and providing them with the tools and technologies they need to navigate the challenges they face. In every sensor designed, in every image captured, there's a silent promise—to support those who serve, and to contribute, in my own way, to a safer and more secure world.

Lessons Learned

Throughout my career, the tapestry of lessons woven has been rich and transformative. From navigating the tumultuous seas of failure to standing steadfast in the authenticity of my journey, each experience has been a brushstroke painting the canvas of my professional narrative. The profound realization that setbacks are not roadblocks but stepping-stones to growth has been a guiding light, teaching me resilience in the face of adversity. Embracing authenticity became a compass, steering me through the male-dominated corridors of engineering, as I discovered the power of being true to myself, even in unconventional choices like the pink lace power blazer. However, it wasn't just about personal triumphs; it was about finding my "why." Whether it was delving into the depths of underwater robotics or inspiring the next generation through The Princess Program for Budding Engineers, I uncovered the essence of purpose in every endeavor. Through facing failure, embracing authenticity, and discovering my why, my career journey has become a testament to the transformative power of resilience, individuality, and a deeply rooted sense of purpose.

Facing Failure

Embarking on the journey of innovation and exploration, I've encountered my fair share of setbacks—moments when the road seemed daunting, and success felt elusive. Yet, it's precisely in these moments of failure that the seeds of growth are sown, and the true essence of resilience is revealed.

Take, for instance, the high-altitude balloon photography project, where we soared to new heights, quite literally. However, the thrill of the ascent was met with the sting of loss as the antenna signal disappeared into the vast expanse, and our payload, along with its precious data and images, never returned. It was a stark reminder that in the pursuit of the unknown, setbacks are inevitable. Yet, within the depths of that disappointment, we found the resilience to regroup, learn, and approach the next venture with renewed determination.

In the realm of my undergraduate senior capstone project, the dream of a self-righting robotic cane proved to be a complex challenge.

Hours were invested in designing intricate parts, only to face the reality that machining them within the given timeframe was beyond our reach. As the project deadline loomed, we found ourselves presenting not a fully realized vision but the base of a cane—a humbling experience that taught me the importance of adaptability and the art of embracing imperfection.

Then, there was the endeavor in Japan, where a seemingly innocuous decision to use fishing weights as ballast turned into an unexpected lesson. The exposed pins of my Arduino board, placed carelessly atop metal weights, resulted in a fried circuit. Yet, within the disappointment, a phoenix of knowledge rose as I reconstructed parts of my experimental setup. It was a vivid illustration that even in the face of unintended consequences, there lies an opportunity for innovation and improvement.

Failure, far from being a roadblock, becomes a steppingstone for growth. It teaches us to pivot, to reassess, and to approach challenges with newfound wisdom. Each setback is a brushstroke on the canvas of experience, creating a masterpiece of resilience and determination.

So, to those facing the sting of failure, know that within those moments lie the seeds of your future success. Embrace the setbacks, learn from them, and let them propel you toward new heights. In the tapestry of innovation, every snag and stitch contribute to the beauty of the final masterpiece.

Embracing Authenticity

In the tapestry of my journey through the male-dominated landscape of engineering, I discovered the power of authenticity and resilience—lessons that would shape not only my career but also ignite a passion for inspiring the next generation of trailblazers.

It all began with a seemingly audacious choice—the pink lace power blazer that I proudly wore to a job fair. In a world where grey suits and muted tones often dominated, my choice was a declaration of my authentic self. The initial response from a company, expressing that they didn't typically hire mechanical engineers, might have deterred others. But armed with confidence and determination, I delivered a compelling elevator pitch that left an impression.

Weeks later, the unexpected call came. The company had secured a new contract that demanded their first mechanical engineer, and they remembered the girl in the pink lace power blazer. It was a pivotal moment—a testament to the power of standing out and confidently owning one's unique identity in a field where conformity often reigns.

As my career in engineering unfolded, I discovered an unexpected avenue for personal and professional growth—pageantry. At the age of 24, I ventured into this world not as a pursuit of superficial glamor, but as a deliberate choice to enhance my public speaking skills. Little did I know that this seemingly unrelated venture would become a catalyst for something much more profound.

The realization struck during my pageant journey—I had a platform to make a difference. Instead of conforming to the stereotypes associated with pageantry, I chose to use the spotlight to inspire the next generation of young girls to pursue STEM (Science, Technology, Engineering, and Math) careers. Thus, The Princess Program for Budding Engineers was born—a groundbreaking initiative that provides free resources for parents, educators, and community leaders to teach young girls in grades K through 5 about engineering principles through the magic of Princess Play. The Princess Program transcended the boundaries of conventional STEM outreach programs. By blending the enchantment of fairy tales with fundamental engineering concepts, it aimed not only to educate but to instill a sense of wonder and possibility. As I adorned my crown and embraced the title of a STEM advocate princess, I realized the profound impact of authenticity and empowerment.

Through this initiative, I became a beacon of possibility, challenging the stereotypes that often deter young girls from pursuing careers in science and engineering. It became a journey not just about breaking barriers for myself but about dismantling barriers for those who would come after me.

In the end, my story is not just one of personal triumph but a testament to the transformative power of embracing one's authentic self. It is a narrative that defies stereotypes, inspires the next generation, and proves that in a world dominated by norms, it's the audacious ones—the ones in pink lace power blazers—who truly make history.

Finding My Why

Growing up, I was fortunate to have a beacon of inspiration in my own home—my mom, a formidable force in the engineering world during her tenure. Though she has since pivoted from the field, her impact on me was profound. She not only broke through barriers but also instilled in me the unwavering belief that engineering was not just a career but an avenue where dreams could soar. As I find myself on the precipice of motherhood, the prospect of being a girl mom brings with it a renewed sense of purpose. I want my daughter to grow up in a world where wild and crazy dreams are not just allowed, but encouraged.

Thus, the torch of mentorship has become a guiding light in my journey—a commitment to paying it forward and ensuring that every young girl sees in herself the potential to be an architect of her dreams. Harnessing the power of my doctoral education and professional experience, I embark on a multifaceted approach to mentorship—individually guiding aspiring minds, actively participating in engineering conferences, hosting impactful women in STEM events in my community, and crafting activity guides to empower parents in nurturing their children's STEM education journeys.

Mentorship, to me, is not just a responsibility; it's a privilege—a chance to contribute to a legacy of empowerment and resilience. My own journey was paved with the unwavering support of a female role model, my mom, who demonstrated that glass ceilings were meant to be shattered. As a woman in STEM, I carry the torch forward, striving to create a path where the footsteps of those who follow are imprinted with the belief that they belong, excel, and lead.

The impetus for my commitment to mentorship lies not just in the past but in the future, embodied in the tiny heartbeat within me—a promise of new beginnings and endless possibilities. As an expectant girl mom, I find myself contemplating the world that my daughter will inherit. I envision a landscape where her dreams are not confined by gender norms or societal expectations, where her curiosity is nurtured, and her potential is limitless. Mentorship becomes the vehicle through which I contribute to building that world—a world where young minds, regardless of gender, race, or background, are encouraged to pursue their passions with unbridled enthusiasm.

My approach to mentorship is multifaceted, reflecting a commitment to impact at various levels. Individually, I engage with aspiring minds, offering guidance, sharing insights, and being a supportive presence in their academic and professional journeys. The one-on-one connection fosters an environment where questions are encouraged, doubts are addressed, and dreams are affirmed. It's about being the role model I was fortunate to have—a tangible representation that success in STEM is not an exclusive club but an inclusive realm where diversity strengthens innovation.

Active participation in engineering conferences becomes another avenue to champion the cause of women in STEM. It's not just about attending sessions and presenting research; it's about being a visible and vocal presence in a space that has historically been male-dominated. By sharing my experiences, both triumphs and challenges, I aim to demystify the narrative around women in STEM, inspiring others to see themselves not as outliers, but as essential contributors to the collective progress of the field.

Locally, in my community, I've taken the initiative to host women in STEM events—an opportunity for networking, mentorship, and collective empowerment. These gatherings serve as forums where women in various stages of their STEM careers can share insights, forge connections, and amplify each other's voices. It's about creating a community where shared experiences become a source of strength, and mentorship extends beyond the individual to the collective upliftment of the entire group.

Additionally, I recognize the importance of empowering parents to play an active role in their children's STEM education journeys. To this end, I've been crafting activity guides that offer accessible, hands-on STEM activities for parents to engage with their children. These guides are not just about teaching technical skills; they're about fostering a mindset of curiosity, problem solving, and resilience. By empowering parents, I aim to create a ripple effect, whereby the seeds of STEM interest are planted early, and the next generation grows up with an innate sense of belonging in the world of science and technology.

In essence, mentorship is a dynamic force that transcends individual aspirations—it's about creating a cultural shift. It's about dismantling stereotypes, challenging biases, and reshaping the narrative around women in STEM. As I navigate this journey of mentorship, I am acutely aware that I am not just guiding the next generation; I am actively contributing to the transformation of a culture. It's a culture where a girl can look at a lab coat and see not just an article of clothing but a symbol of possibility. It's a culture where STEM is not just an acronym but a call to action—an invitation to dream, explore, and innovate.

In paying it forward, I am not just investing in the future of STEM; I am investing in a future where every young mind, regardless of their gender, sees themselves as architects of progress. Through mentorship, whether one-on-one or in collective spaces, I aim to be a catalyst for change—a source of inspiration that echoes through generations. As I await the arrival of my daughter, I envision a world where her wildest dreams are not just possible but inevitable—a world where mentorship becomes a legacy, and every girl is empowered to reach for the stars.

Connect with Kelsey:

Empowering the Future: Insights from 30 Years as an Electrical Engineer

Coauthored by Kathy Nelson

Engineering Dreams…?

WHAT IS AN engineer and what does an engineer do? These are not questions I could answer when I was 16, 17, or 18 years old. I couldn't even answer those questions when I was 20 and studying to be an electrical engineer. Knowing what a wide variety of jobs and careers entail, especially ones we don't see up close on a regular basis, can be very difficult. Did I know when I was 17 years old and making decisions about college that I wanted to be an engineer? No, I did not. At that time, I wanted to be an architect. I did make my decision to be an architect very early on, however, and for a very specific reason.

In 1980, when I was between second and third grade, I was fortunate to attend a one-week summer day camp where us kids shadowed professionals. It's astonishing to me now to think that there was a summer camp in 1980 that let little kids shadow professionals. And it gets better: I shadowed a *female* architect! At that time, less than 4% of licensed architects were women.[1] The odds of me shadowing a female architect in elementary school in 1980 are incredibly small. However, it had a serious impact on my life. From that week on, I wanted to be an architect!

The desire to be an architect shaped the classes I took in high school. At the time I was told architects use a lot of math and science, which I loved, so I took a lot of those classes and also took whatever drafting classes I could in high school, as I was told there was a lot of drafting required in architecture.

When it came time to apply for college, I applied for and was accepted into architecture school. I was on my way! I thought. But then I found most of my classes to be art-related! I had been told that architecture was full of math and science. I loved math and science! My classes did not use a lot of math and science, though. I didn't know what else to do for a career. What could I major in if I wasn't going

1 I was unable to find data specific to the year 1980, but national surveys cited in an American Institute of Architects web article "Women in Architecture" showed that in 1988 only 4% of registered architects were women, up from only 1% in 1958 (https://www.aia.org/articles/6252982-women-in-architecture).

to be an architect? What did I know other than that I loved math and science?

Fortunately for me, my dad was an Engineering Technology Professor and had always wanted one of his daughters to be an engineer. My two older sisters had opted for other career paths; as my father's youngest daughter, I benefitted from his support and encouragement when I began to look for a different major and career. I didn't know what an engineer was, but I knew it used math and science. My new brother-in-law had recently started work as an electrical engineer in the Air Force and he seemed to like his job.

At the time, there was no internet, so researching jobs and careers was difficult. With that minimal knowledge about engineering, I changed majors. It's amazing to me how little I knew when I made this decision about what would become my life's work, but after 30 years of being in the industry and talking with hundreds of women in STEM (Science, Technology, Engineering, and Math), I don't think I'm alone. Many of us don't know what careers and jobs even exist, or at least what they entail on a daily basis--I think that's very difficult to know from the outside. We make the best decision we can at the time and hope it's the right one. Or at least a good one. What started out for me as an unknown career path in electrical engineering, when I was a college sophomore, became a satisfying career that I have thoroughly enjoyed.

My engineering classes at North Dakota State University were challenging and usually fun. During my junior year, I took a power systems course with an energetic and engaging professor, Dr. Stuehm. He was lively and animated. He made power systems seem fun and exciting, even though we never actually saw substations or electrical equipment. It's interesting how much of an impact a teacher can have on our career choices. Because I liked Dr. Stuehm, as did many of my classmates, I continued taking power systems courses and graduated with a Bachelor of Science Degree in Electrical Engineering, with a Power System emphasis. With this type of degree, it is common to work for electric utilities. I was unwittingly forming my career path. I was one of only four women in a graduating class of 75 electrical engineers.

Becoming a Utility Engineer

When I graduated, the job market was not great. Also, unfortunately, I hadn't done an internship or cooperative[2] and had no job experience. I struggled to get interviews, let alone land a job. Well into the summer after graduation, I was lucky to get an interview with an electric utility in Minnesota, United Power Association--not for a job as an engineer, but for a job as a drafter. By this time, I was desperate for any job, and I had experience in drafting both in high school and at a summer job where I worked as a drafter for a small engineering firm. My interview for this drafting position was another career shaping moment for me.

My interview for this drafting position was with Jim Goodin, the manager of telecommunications. I had hated my communications courses in college on account of a sexist and uninteresting college professor (again, the influence teachers can have, both good and bad!), so I had no interest in communications. However, Jim made a comment during the interview that changed my perspective and my career. He said offhandedly, "Sixty cycles (this is how he referred to the power system) has been the same for a hundred years. Telecommunications is always changing and is the most interesting place to work in the utility industry." Whether it was his comment or the need for a job that caused me to accept this position, this is how I began my career in telecommunications in the utility industry. Additionally, Jim had assured me there would be an open position for a telecommunications engineer within the coming year, and that I would not be a drafter for long.

The year I spent as a drafter provided a foundation for my future work as an engineer. I learned to be detailed in the drawings I created, and learned the importance of the detail provided to the technicians who were on the receiving end of the drawings I created. I spent much of the first six months of my job in the field with technicians learning about the work they do and the importance of good engineering work

2 Cooperatives are usually full-time employment over the course of a semester. A student would take time off from attending college to work at a company and then return to complete their education.

order packages.[3] Because my utility was union, I wasn't able to actually work on equipment, so if I was in the field and there was a wrong or missing part, I was the one who would be sent to town (which in rural Minnesota might be an hour or more away) to pick up a bolt or screw or something else that was available from the hardware store. This experience helped me learn the importance of making sure all parts are included in a materials list, even seemingly unimportant nuts or bolts, and of ensuring that the proper size of these parts gets detailed. A missing or incorrectly designated part can cause an additional half day or more delay to a project, which increases the budget due to increased work time. These early learning experiences were critical in my development as an engineer, and I have never forgotten them.

Fitting In

My early days as a telecommunications engineer quickly passed. I developed friendships with fellow engineers and technicians. I didn't always feel like I fit in, though, because I was typically the only female in meetings, in my department, and in the field. There was one other female engineer at the company, a power engineer who worked mostly in the office, running transmission capacity studies. She and I became fast friends as we both had moved to the small town our utility was located in when we were just out of college, and we were the only women engineers at the company. Her friendship was incredibly important to me, and she is still one of my best friends to this day. Forging friendships and community with other women in our field is incredibly important and I am ever grateful for her presence in my life.

One of my other learning experiences as a young engineer is a bit dual-edged. As the only woman in my department, and especially because I was often going out into the field with male technicians, I wanted to "fit in with the guys." At the time, this seemed necessary. And so I downplayed anything I did that was feminine. I sewed a lot of my own clothes and enjoyed baking, especially baking and decorating fancy cakes. I never talked about these hobbies at work, because I didn't

3 A work order package is a set of instructions, material list, and drawings that is sent from engineers to technicians who install and commission equipment.

want to be seen as "girly." I tended to hide these sides of my personality—a perceived necessity which I now find unfortunate. However, I'm not sure I would do anything differently. This was 30 years ago. Times were different.

I also started golfing when I started working at this utility. I did this with the intention of being able to network with engineers and other men at work. I wanted to have something I could do socially with colleagues and coworkers. Golfing to fit in at work has become a bit of a point of contention for many women recently. I see a lot of posts debating women golfing or not golfing and questioning golf networking events. Women post questions such as "If I don't golf, am I being excluded from discussions and decisions that are being made?" There are valid points on all sides of the issue, and I don't have the answer. For me, 30 years ago, I wanted to have something I could do with the people I worked with, so I learned to golf.

My golf experience has been great. I formed a group of women golfers at my utility, and we golfed every week during the summer (until I had kids, when that became unsustainable). When I would travel overnight to a job site with technicians, we would often golf after work. These were great experiences for me. Many of my fellow engineers struggled with having good working relationships with the technicians. There is sometimes some tension between technicians and engineers. I developed great relationships with many of the technicians I worked with because of the time we spent golfing. I got to know them on a more personal level, and we'd talk about families and personal lives while we were on the golf course.

For me, golf allowed me to connect with co-workers and colleagues throughout my career in a way I wouldn't have otherwise. While I am not a great golfer, I have been able to hold my own. My drives aren't long, but they usually are straight, which typically means I can keep up with my foursome (and it helps to use the women's tees which are usually a little closer to the greens).

Career Progression and Family

As I moved from an entry level engineer to mid-career, I was given more responsibility. Within five years of starting my job, I was given the responsibility of managing a $30 million dollar project. Whether this was because I worked in a smaller company or because my manager thought it would be good to give me a large responsibility early on, I don't know. What I do know is that I am grateful for that opportunity because it shaped me and gave me an incredible experience at a young age. At that time, there were not Project Managers as a separate job function. Engineers were project managers, technical experts, and researchers. They staffed projects, developed budgets, and more. I learned more in the two to three years of managing this large project than I ever expected to in the early part of my career.

While I started my first major project, my husband and I decided it was time to start a family. Great timing: large project and pregnancy. I was the first woman my boss had ever managed. He had to find out about maternity leave policies and how our company dealt with pregnancies, return to work after leave, and more. Maternity leaves were pretty short, typically only six weeks, and paternity leaves were years away – my husband got only three days off when our daughter was born. As I was at the start of a large project, I timed my maternity leave to be right after the Request for Proposal was released for this project. This meant vendors had ample time for responses and there was not a huge pile of work waiting for me when I came back from leave.

Coming back to work was hard. Leaving a six-week-old baby in the care of someone else is hard. Deciding to continue working or to be a stay-at-home mom is hard. People can be very critical of others. Everyone has an opinion. At that time, there were not lactation rooms at companies. Pumping was typically done in a bathroom stall. I remember one day I had a meeting on the other side of the state with my boss. This was an important meeting with a large group of people. My boss asked me what time I wanted to leave. I said we should leave by X time, or we could leave a little later, but that would mean I'd have to pump on the way. He said "okay." I'm not sure he actually understood what I meant when I said that or if he didn't care, but I ended up pumping on the way in the back seat of the car. While this sounds inappropriate and maybe was, it makes for a funny story now. I was pretty good friends with my boss and sometimes you just have to go with the flow and find humor in situations. I doubt that I would have made that same choice now, but at the time, that's what I did.

Navigating New Career Opportunities

As my career went on, I began to look for new challenges. Within seven years, my husband and I had had three babies and life was busy, but I also loved working as an engineer. We moved to a town about two hours from the office I worked at, and I began working remotely three days per week. At the time, remote work was not common, and working remotely stymied my ability to move into leadership at my company. I had progressed to Principal Engineer, the highest technical level I could go. Around that time, I met the Chairman of the Board of a trade association whose conferences I had attended from time to time over the years, Utilities Telecommunications Council (UTC).[4] He suggested I consider joining their national board of directors. I was looking for a new challenge and was intrigued.

I met with the Chair of the Technical Division at the next annual national conference, and we decided it was a good fit. My company was supportive and, because I said "yes" to this opportunity, I started on another path that would be career-changing. I just didn't know it yet. When I joined the board of UTC, there were about 80 board members. I was one of only two women on the board. As usual, I found myself a very small minority, but I was used to it and I loved the board of UTC. I was welcomed to the board by a group of older utility men who ended up becoming some of my best career friends.

Once again, I found myself trying to be "one of the guys"--this time a little less successfully. Cigar parties are a linchpin at UTC conferences. However, I don't smoke cigars and never went. The cigar parties are one of the things that made me think UTC was a "good old boys" club that I didn't want anything to do with, other than to keep up my continuing education credits for my professional engineer's license,

4 Now Utilities Technology Council.

and to keep up-to-date on technology. But now that I was on the board and many of the board members were smoking cigars at the end of the evening, I decided to try--so I smoked my first, and last, cigar my first night as a board member.

Now, I had learned how to golf to "fit in with the guys," so I figured I could smoke a cigar – how hard could it be? I had many "coaches" telling me to make sure I didn't inhale. So, I didn't inhale. I smoked my cigar and proceeded back to my hotel room late that evening--evenings tend to be late at a UTC board meeting. I got back to my room, lied down on my bed, and immediately threw up on the floor! Not once, but twice--barely missing my laptop which was laying next to the bed.

The next morning, we had to be up early. Board meetings start early, and on this particular day we had board meeting guests from Capitol Hill coming in, which is typical for UTC board meetings. Board meetings, especially days with Congressmembers present, require business formal dress, typically a suit. I was still feeling a little green from the night before as I came down for breakfast to be greeted by my fellow board members with "Kathy! How are you feeling? You're looking a little pale." To which I responded: "I got a little sick last night." And their response? "Oh, that always happens with the first cigar." Seriously?? Could someone have mentioned that the night before?? I might have done things a little differently! I now am solely a spectator when I attend cigar parties or cigar-smoking events and, 15 years later, I still hear jokes made about that experience – all in good fun.

For me, UTC was a place where I found my voice. At board meetings we met with people in Washington D.C., and talked about the issues and challenges utilities face with policy and regulation. I spoke up at these meetings and was encouraged by my fellow board members.

A few years into my tenure on the board, the Chair of the Public Policy Division (PPD) was going to be retiring from his utility and stepping down from the board. I had gotten to know him pretty well—we initially broke the ice when he walked with me the day after the cigar "incident" to buy Febreeze from the local convenience store to clean my carpet. He asked me if I would be interested in taking over his role as Chair of PPD. At the time I had little interest in politics, but he assured me this really wasn't politics, it was continuing to use my voice to advocate for utilities telecommunications' interests. Again, I said "yes."

I loved chairing PPD. During the time I chaired that division, UTC was advocating for specific spectrum requests, utility funding for broadband, access to FirstNet, and more. I went to Washington D.C. five or six times a year to meet with the Federal Communications Commission (FCC), National Telecommunications and Information Association (NTIA), Department of Energy (DOE), FirstNet, and for Board Days[5] at Capitol Hill. I really enjoyed advocacy and sharing with lawmakers the challenges that utilities face. This is not to say it was easy or always went our way. Mostly, it didn't. But when I neared my three-year term as PPD chair, several people approached me asking if I would be interested in moving into UTC board leadership. There is a very structured path going into board leadership at UTC. The first year you are secretary/treasurer. The second year Vice Chair. The third year is Chair. The fourth year is Immediate Past Chair. It's a long commitment. I was excited about the opportunity, and the support and encouragement I had from the UTC board was incredible. I felt valued. I felt like I belonged. And I felt like I could make a difference. So again, I said "yes."

The Power of "Yes"

The year I chaired UTC was the most enjoyable year of my career to date. I had an amazing experience. I was the first Chairwoman of UTC, a 70-year-old trade association at the time. At the meeting where I was elected, in a room full of well over 500 people, I was welcomed to the stage to accept the nomination with a standing ovation. I had never before received a standing ovation--it's an incredible feeling! It's a feeling of belonging and validation that I hadn't experienced before, and it felt great.

As I began my acceptance speech, a speech I had been preparing

5 Board Days are days the UTC Board meets with many senators and members of Congress to discuss utility telecommunications and technology issues.

over the past three to four months, the electricity in the entire ballroom went out. No lights, no microphone; nothing. Now, this is a utility trade association in a room filled with mostly utility workers, so the irony of this was not lost on anyone. Normally, this would have been cause for stress and panic. However, I had prepared well for this speech and worked with a communications coach, so the 5-minute break while they found and fixed the issue actually calmed my nerves. Many of my board member friends were at the front of the room and I chatted casually with them while the issue was fixed. The lights and microphone came back on, and I continued with my speech.

As an employee of a medium-sized electric cooperative in Minnesota, I did not have the opportunity to travel internationally for work. The year I chaired UTC, I got to travel internationally. My first trip as the new Chairwoman for UTC was to Cape Town, South Africa, for Africa UTC and Africa Utilities Week. While Africa UTC was small, it was hosted in conjunction with Africa Utilities Week, which was huge. One of my favorite experiences at Africa Utilities Week was a "Women in Power and Water" luncheon. At the time, I held a stupid stereotype in my head about women in Africa, so I was not prepared to walk into the ballroom of this luncheon to a room of 100 or more women engineers. This was not anything I had experienced even in the United States at the time – I'm not sure I have yet, six years later. It was incredible.

The keynote speaker was the Minister of Water and Sanitation of Uganda. She has a PhD in Chemistry. I also met many women engineers from around Africa both at that luncheon and throughout the conference. I had incredible conversations and I learned that most of my stereotypes about Africa were wrong and unfounded. I realized that I needed to start checking stereotypes that I held. Honestly, my ignorance on this score was pretty embarrassing. This trip to Africa was life-changing for me in many ways, only one of them being career-related.

Later that year, I traveled to Lisbon, Portugal for European UTC, and Salvador, Brazil for UTC America Latina. As Chairwoman of UTC, I delivered keynote speeches at these conferences. During the year, I was also invited to deliver a keynote speech to a group of

Women in Energy in Indiana. This was the first time I was invited specifically to deliver a keynote speech, another incredible experience. I was a shy, quiet person in my early years, but as I've aged and gained confidence, I've found that I like to speak in public. Initially, it was scary, and it always got my adrenaline pumping. I continued to work with a communications coach, though, and I learned that preparation was key. I had fun and I found I could connect with my audience. I also think people liked listening to me. This was the first time I started giving presentations that were not technical, and I had to develop a new skill: owning a stage.

I developed many new skills during the year I chaired UTC: leadership skills, financial management skills, managing a non-profit board of directors, dealing with controversy (there was a controversial issue that came up the year I chaired), prioritizing strategic imperatives, and listening to people and making them feel heard and valued. I met incredible people across the United States and around the world.

Most importantly, my days on the UTC board taught me the importance of saying "yes" to opportunities when they arise. They also taught me to seek out opportunities in case someone doesn't come knocking on your door presenting you with them. If you don't seek them out, you may miss out on a fabulous experience. I also learned that things are not always what they seem from the outside. We need to check our biases and question our stereotypes. I loved the years I spent on the UTC Board. I made lifelong friends and built the most incredible network. And I had fun!

A Time for Change

In mid-2018, after completing my term as UTC Chairwoman, I was recruited by a small radio manufacturing company (Ondas) by people I had known for years. They were growing rapidly and asked me to join them to help grow the company and to work on an IEEE standard[6] I had helped them launch during my time at Great River

6 IEEE, the Institute for Electrical and Electronics Engineers, is "the world's largest technical professional organization dedicated to advancing technology for the benefit of humanity" (https://www.ieee.org).

Energy.[7] Leaving my utility of 25 years and the only career home I knew was one of the most difficult decisions I have ever had to make. However, I had limited growth opportunities at GRE due to working remotely, and I had a new manager who I struggled to work with. They say people don't leave companies; they leave bosses. That definitely was a contributing factor in my decision, so I said goodbye to my co-workers, resigned from the Board of UTC (I was Immediate Past Chair at the time) and moved on.

My role at Ondas was as the Director of Technical Product Marketing and Industry Relations. In this role, I spent a lot of time educating mission critical industries such as utilities, energy, transportation, and rail on an IEEE standard we were initiating through the IEEE standards body. I used many of the skills I learned at UTC to deliver a lot of presentations and webinars, and I wrote white papers.[8] I spent the majority of my time communicating, both verbally and in writing. (Note to any students or young people out there: communications are important! Take those speech and writing classes!!)

I also got involved in the IEEE standardization process. Another new and interesting endeavor! The standards body I was involved in was IEEE 802. This is the standards body responsible for Ethernet, Wi-Fi, and more. While I frequently felt like a minority as a woman at conferences throughout my career, I was even more of a minority in IEEE standard meetings. It was rare to find even one woman in the standards meetings I was in. I hope that is changing, but there was definitely a significant disparity at the time. The IEEE 802 meetings are huge. Probably 1,000 people. I may have seen 20 women total during my involvement! However, it was a fun learning experience. I hadn't been involved in standards development before and I got to learn about the process and work to bring people together in different mission critical industries to support the standard.

---

7 Great River Energy was the product of a merger of United Power Association and Cooperative Power Association in 1999.

8 A white paper is "an authoritative, research-based document that presents information, expert analysis, and an organization or author's insight into a topic or solution to a problem. Companies or vendors use these papers in business-to-business (B2B) marketing models as part of a content marketing strategy." (Nick Barney, "What is a White Paper?", techtarget.com)

I also got to travel, which I loved. The company I worked for was located in Santa Clara, California, so trips to the office were to California, which was fun. The IEEE meeting I attended was in Kona, Hawaii, which is a great place to visit. I joined trade associations and learned about new industries.

The company I worked for became very involved in the rail industry. It was fascinating for me to learn new industries and come to understand their challenges. There was much that was similar to the utility industry: the need for mission critical communications, as an example, and much that was not: unlike trains, substations do not travel at 60 miles per hour while they communicate. I was also part of a company that was growing at a fast pace, which was exciting.

A Global Pandemic and My New Passion

In the spring of 2020, shortly after the pandemic hit, my company shut down temporarily and all employees were furloughed. Life took an unexpected turn and I found myself out of work, at least temporarily. I hadn't been unemployed since I was 15 years old. I had two of my three kids still at home (one was off at college), but they were older and didn't need constant care and attention. I had time on my hands. The idea of hosting a podcast had been seeded in my mind months before, while at a conference with a lunch speaker who hosted a podcast. As part of her talk, she mentioned the idea that people should consider hosting podcasts. It wasn't something I had strongly considered, but the seed was planted.

Around this time, I was an avid listener of Dax Shepard's "Armchair Expert" podcast where he did long-version, casual conversations with actors and other well-known people, or on interesting subjects. He had several actresses on his podcast who would talk about the challenges of being female in Hollywood, especially as they tried to move into directing and writing and away from acting. When I would hear these stories, I would think, "This sounds like every woman in STEM I know!" yet the only woman we heard from at that time was Sheryl Sandberg, the COO of Facebook and author of "Lean In." I decided to try to start a podcast.

Starting a podcast seemed daunting, but it turns out it's really not that hard to do. I researched what equipment to buy: all I really needed was a good microphone and an average set of headphones. I had a Mac computer that came with GarageBand,[9] an audio editing software application. The only other thing I needed was a podcast recording platform, and later a podcast hosting platform. Zoom became prevalent during the pandemic, so I bought a subscription so I could record longer than 40-minute episodes. It also allows for separate audio tracks to be recorded.

I convinced my two engineering besties, Kerry and Tami, to be my first two guests for Episode 1 and 2, then sent the episodes to a number of people to get their feedback. People liked it! I recruited eight more people to interview and launched my podcast with 10 episodes in June 2020. There were a few other business logistical things I had to figure out. I had to find a name, for example. I researched names that were already being used and tried to trademark the name I had chosen. I also bought a domain name. I built a website. I even had merchandise made and opened an Etsy store. Although this merchandizing foray was unsuccessful, it too was another lesson learned.

I loved my podcast. What started out as interviews with women I knew, soon branched out to include women who were friends of friends. Eventually strangers were reaching out asking to be on my podcast! I met AMAZING women! They are so smart, so inspiring. My podcast quickly became my passion project. I began thinking about the impact and importance of encouraging and supporting girls and women in STEM again, something that had been on the back burner for me during the busy years of raising kids.

Editing podcasts was another matter. It was extremely time-consuming. I spent A LOT of time on my podcast. Editing a podcast episode took about eight hours, so with the research, recording, and editing I was spending about 12 hours per episode and releasing an episode per week. This was fine during COVID as we were stuck at

9 GarageBand is a line of digital audio workstations developed by Apple for macOS, iPadOS, and iOS devices that allows users to create music or podcasts. (Wikipedia)

home, so I had the time. As we started coming out of COVID, though, I had to make some changes.

I changed jobs during the summer of 2020. My heart was no longer at my current company. After leaving Great River Energy, which had been a very difficult decision, leaving this job was easier. Job changes no longer had the gravity I once thought they did. There are more jobs out there and each job brings a new set of skills, more people to meet, and great opportunities.

I had moved onto consulting and joined a firm based in Chicago. Thanks to COVID and the increase in remote working opportunities, I had the ability to work in places I may not have otherwise, since I live in a very rural area of Minnesota with very few opportunities as an electrical engineer. Consulting is very busy. My days quickly became full of meetings and work. The eight-hour workday of utility work was long gone and finding time to continue my podcast weekly became almost impossible. With a niche podcast, I have not been able to monetize podcasting and couldn't afford an editor, so I continued to do all the work myself. I reduced my podcasts to every other week, and even then, struggled to keep up with it at times. My husband complained that I was on my computer all the time--which indeed I was. I edited podcasts nearly every night after work. At the same time, though, I really love my podcast and the inspiration it brings me. I also get emails from women periodically about how they love the podcast.

One email in particular was from a woman who worked as the only woman engineer at her company. She told me how lonely and isolated she had felt until she found my podcast. I needed to keep going! I needed to provide a platform for women to share their stories. I needed to build a community. So, I persisted. As of the writing of this chapter, I am up to about 120 episodes! Thanks to the women (and the few men) who have entrusted me with their stories. They have made my heart so happy over the years and touched the lives of countless women all over the world.

Unexpected Changes

After three years as a consultant, I was abruptly laid off in the summer of 2023. It was the week after the UTC Annual Conference, which is like a family reunion for me, and I had been feeling on top of the world: giving presentations, making connections, and building relationships that would hopefully become business for the company I worked for. The economy had been growing weaker and I knew other areas of the company were struggling. There had been layoffs earlier in the year, but they hadn't affected our part of the company. We still had a lot of work and had been told we were fine. I awoke one morning to an email saying there were layoffs, but if you didn't have a one-to-one meeting scheduled that day, you were not affected. I closed the email thinking about the poor people who would be being laid off and felt bad for them. Then I saw my one-to-one meeting scheduled, and my heart dropped. It was very unexpected; I thought I had been doing all the right things. I won't go into detail here, but it was hard. I had friends, both coworkers and clients, who I loved working with. I spent the day in shock, saying goodbye to people and trying to figure out what to do in the few hours before my computer and network access would be gone. It was just very hard.

If one is going to be laid off, it's helpful to have a mission trip planned. Two days after being laid off, I boarded a plane to St. Croix with my husband and son to spend a week on a group mission trip with youth groups from around the country. We had had this trip planned for several months and it couldn't have been better timed. It's hard to wallow in self-pity when you are serving other people in significantly more challenging circumstances than you are in. I also had three priests and a pastor around most of the week who I could talk to about my situation. I'm a pretty faith-filled person, so trying to understand what God had in His plan for me and grounding myself in that was incredibly helpful. I also met two women chaperones who were near my age who had recently lost their husbands very suddenly. I spent a lot of time talking with them and listening to them share their memories with me. My job loss began to seem inconsequential in comparison to their losses. I would be fine.

For the past couple of years before my layoff, I had thought about consulting on my own. While I loved my coworkers and clients, I didn't always love the corporate world. There was competition, pettiness, and internal politics. After 30 years of corporate work, it was getting harder to deal with some of those aspects. Starting my own consulting company was a natural fit. While I did try to take a month off after being laid off to "not think about what was next," I was always thinking about it. I had to. I contribute very significantly to our household budget. We can't afford our mortgage or lifestyle without my paycheck. Barely over a month after being laid off, I filed the paperwork and began my consulting company. I had a couple of small projects right away--I am ever grateful for my network and the people who gave me work early on when I was just starting out. It is one of the kindest things someone could do for me at that time. It's hard to be laid off, and even though I was part of a mass layoff, my confidence had taken a hit. Having friends reach out with work helped rebuild that confidence.

I am now in a new chapter of life and work. I am an entrepreneur. I own my own business. I set my own schedule (for the most part, as I do have clients, and they have some impact on my schedule). I am my own boss. It can also be a bit stressful. I have to find my own work. I don't have a steady paycheck. Sometimes I have too much work. I love it. I do miss my coworkers and have found I need to be mindful to reach out to people to connect. I have to plan social activities to ensure I am taking care of myself mentally, because I talk to less people on a daily basis than I used to. I have a great network.

Thanks to my time on the Board of UTC and to my podcast, I have friends all over the world who I can connect with. My podcast also grew my network and I have been going to virtual networking events since COVID. I learned long ago the power of connections and the importance of community. These friends I have made over the years have been my lifeline lately. They are my village. Some of them I've never even met in person…yet.

A Life Well Lived

After 30 years as an engineer and after going through some job transitions, I have been looking back on my career and life and reflecting. I have had a great career. Every part of my career, both wanted and unwanted, expected and unexpected, has led me to where I am today and provided opportunities to learn and grow. Growth that is unplanned and unexpected is hard, there's no denying that. But I think the times of most growth come from difficult circumstances and unexpected opportunities. I am grateful for where my career has taken me. Most of all I am grateful for the people it has brought into my life. While we all like to have work that challenges us, in the end it's the people, the relationships, and the community that bring fulfillment. I love and am ever grateful for my village.

So enjoy your job. Enjoy your families and friends. Embrace life. Say "yes." Take chances. Take advantage of opportunities that come your way. Be yourself. Be authentic. Lift each other up. Build your village.

Connect with Kathy:

Soaring Against the Odds

Coauthored by Jasmine LeFlore

San Diego County Engineers' Outstanding Engineering Service; San Diego Business Journal's 50 Most Influential Women in Technology in San Diego; Black Engineer of The Year Science Spectrum Trailblazer; and San Diego Magazine's 2022 Women Builder of the Year. These are just a few of the awards I've received based on my work. Today I am proud to say I am the co-founder and executive director of a nonprofit organization called Greater Than Tech, that has served over 1,000 students by teaching them STEM (Science, Technology, Engineering, and Math), as well as entrepreneurship to become the next generation of innovative leaders. I also work for a Fortune 100 aerospace company as an advanced technology solutions lead, where I develop solutions to manage cutting edge technologies projects for industry technical leaders. I travel for work and for fun. I own my own condo in America's finest city (San Diego), and I have an active lifestyle. Overall, I'm proud to say that I am living a life I could have only dreamed of when I was a kid.

I vividly remember the days I used to long for a better life. Growing up in Flint, Michigan along the flight path, I often found myself gazing at the airplanes soaring overhead, wondering, "How do planes stay in the air?" When I asked my mother this question, instead of giving me the answer, she just said, "maybe one day you'll have a job where you'll know how to answer that question." Little did I know that this simple question would set the course for my journey, and shape my aspirations and ambitions.

At a young age, I often remember my family referring to me as "the puzzle wizard" or "jazz-o-matic." I filled my free time doing puzzles with my grandmother, making up board games, creating crafts, and taking things apart and putting them back together. And although I excelled in the classroom, I was verbally shy. My kindergarten teacher noticed that I would whisper answers to friends instead of speaking aloud. Based on my teacher's recommendation to get me out of my shell, my mother enrolled me in diverse activities to shape my social acumen—ballet, tap dance, hip hop, praise dance, church choir, modeling, and more. Soon I was in multiple activities at once, which set the trajectory for how I kept pace with a lot of things at the same time. I liked school, I liked doing all of these activities, and I was a pretty happy child.

However, life can have a way of introducing unexpected twists. My parents divorced when I was nine and everything about my life changed after that. I went to a new school and had far and few visits with my dad, and then the extracurricular activities eventually ceased. The financial stability of our home suffered to the point where my mom and I were evicted several times, and we even lived in a women's shelter for a while. Around this time, I just yearned for a life with normalcy. Amidst these challenges, a glimmer of hope emerged.

Fast forward, to my 8th grade summer, I was a part of the Carrera Program, an initiative designed to prevent teenage pregnancy for at-risk inner-city youth. We went on college tours, and the tour that literally changed my life was visiting the aerospace department at The University of Michigan (U of M). The experience was transformative; standing in the presence of a wind tunnel designed to copy the actions of objects in flight was amazing to me. I already had a fascination with airplanes as a kid, but this tour was my 'a-ha moment.' I knew right then and there that I wanted to study aerospace engineering at the University of Michigan. Then I would be able to answer my childhood question about flight while providing a way to build the life I aspired to. I was truly taken aback when I learned that the University of Michigan held the top position for aerospace engineering programs in America (at that time).

While initially intimidated, I harnessed the shock into motivation. If I wanted to study with the best, I needed to become the best candidate possible. What followed was a whirlwind personal transformation. The smart but complacent student vanished and was replaced by a well-rounded, high achieving all-star. Over the next couple of years, I became the Captain of the Cheer Team, a Track and Field school record breaker, Student Council Representative, a member of the Robotics Team (shout out to Team 397 Knight Riders), and National Honor Society Vice President, all while maintaining a part-time job at McDonalds, just to save up for my car, my senior pictures, and my grad party.

By 11th grade, I was in full swing, I took an engineering graphics class which landed me an internship at Wade Trim Civil Engineering. I even did a pre-college engineering program at Kettering University called AIM (Academically Interested Minorities), where I stayed on Kettering's campus for 4 weeks, and took engineering classes like a real college student. These opportunities to have hands-on experience with STEM were like dipping my toe into what my real life as an engineer could be. As if my schedule wasn't jam-packed enough already, I doubled down cramming for the ACT, taking it again and again - refusing to quit until I achieved a suitable score.

With my sights set on the University of Michigan, I faced skepticism from an unexpected source – my high school counselor. When I expressed my desire to attend this prestigious institution, he laughed dismissively, suggesting that I wasn't "University of Michigan material." The words stung, but they ignited a fire within me ... and also more questions. I remember thinking ... maybe he's right. Maybe I should check out these other schools on my list of pros and cons. I remember showing my high school physics teacher all of the schools I was considering, and he pointed to U of M. He told me, "You belong there."

Despite my fierce preparation for my college application to the College of Engineering at (U of M), I was denied entrance. As you can imagine, I was crushed. All of my hard work and effort for the past 3 years seemed as though it meant nothing. But luckily all hope wasn't lost, I was accepted into the College of Literature, Science and Arts (LSA) at U of M. My then-admissions counselor recommended that I attempt a cross-campus transfer, where I would take pre-requisite classes to get into the engineering college. However, when I got to U of M, I also had to do a summer bridge program.

Through this intensive prerequisite curriculum, I confronted harsh realities that my high school foundation left me underprepared for the cutthroat demands of excelling at such an elite institution. Another conversation that burned in my memory was one with my first college advisor. She told me I'd never get into the college of engineering since my ACT score was lower than the average applicant. She reminded me that I came from an inner-city school and I was black and female, so I should just be grateful that I was even admitted into U of M. (At the time, Black students at U of M only represented 4% of the 50,000-student population.) She even told me I should get a degree in general studies, just have fun, and be a manager at Target. I told her I'd rather flunk out trying to achieve a goal that I wanted before I got a degree in something I had no interest in.

The normal process for a cross-campus transfer is 2 years, but it took me almost four years to complete. I was clearly not doing well in my classes. My plate was still full - I was working two jobs, sending money to my mom and my brother, and ultimately, not having fun. I felt like I was wasting so much time trying to do something that just simply did not seem attainable. Naturally, I was questioning if studying to be an engineer was worth the risk, and wondering if my college advisor and my high school counselor were right. I was destined to find out.

I spoke to the department chair of the aerospace program about wanting to transfer and he gave me a waiver to just take an aerospace class and see how I did. Luckily, I was successful with the class, but it didn't really balance out the other classes that I wasn't doing well in. I was still over my head and trying to stay afloat.

On top of all my responsibilities, what made matters worse was that I experienced tragic loss when my mom unexpectedly passed away in my junior year of college. A month later my grandfather passed away, followed by my grandmother eight days after him. These were the worst things that ever happened to me. It felt like my world shattered. But what was I going to do, just drop out? I was the first in my family to go away to college. I knew they were proud of me for making it this far. I did take a leave of absence from school, but all could remember was the echo of one of my grandad's commonly said phrases .. "A winner never did quit, and a quitter never did win." At this point, I put all my emotion and pain back into my schooling and I realized that keeping busy helped me cope.

After a grueling four-year journey, I finally secured my prized engineering school admission. But the celebration was short-lived. I still needed to graduate and land an engineering job for this degree to mean anything. With nothing to lose, I snuck into an invite only event to get my first engineering job. I remember the first recruiter

I met seemed confused to see me there because my GPA was below what was required. I asked to be introduced to the hiring manager which the recruiter was reluctant to do, and next thing I knew, I had an interview for the next day. This turned out to be my first engineering job! I was soon moving from Ann Arbor to Hartford, Connecticut to be a component integrated product team lead (CIPT) for United Technologies (now called RTX).

I spent the beginning years of my career feeling like I wasn't meant to be in the work environment, that I was in. Partly because I felt that I didn't technically meet the standards to get in, and partly because it seemed like people were so interested in me because I was different than your typical engineer. Therefore, I was trying to stay under the radar. I would get questions about why I was working so hard, and internally I was thinking because I'm technically not supposed to be here. I finally stopped feeling like an imposter about two years ago when I realized my adversities are what really fueled my tenacity. I feel that my adversities are contributors to my greatest strengths, and I am no longer afraid to stand out.

It's clear that my path has been filled with twists and turns, yet through it all, I persevered thanks to the foundation instilled in me as a child - the encouragement that I could achieve anything I set my mind to. No barrier or setback is too great when armed with self-belief, grit, and the support of those who care about you. Though I've reached heights I only dreamed possible back when I gazed skyward from humble roots in Flint, Michigan, in a lot of ways I'm still that same curious girl asking, "how do planes fly?" Except now I'm helping design them, and now I'm the one asking how I can help the next generation soar.

I hope my story continues to inspire the next generation of young dreamers, thinkers, and doers who have faced doubt in their abilities. As the first in my family to soar to new horizons, I aim to pave the way for others to climb higher on my shoulders. Though the skies ahead hold exciting possibilities, I'll always smile looking back and seeing all the hands I pulled up with me knowing that my impact on others was only possible because I didn't give up on myself.

Connect with Jasmine:

From Zero to 100: Let Your Life STEM From Death

Coauthored by Regan Tracy

My name is Regan Tracy. I'm a 23-year-old senior engineering student with a borderline-crippling caffeine addiction. I'm majoring in aerospace engineering and dual minoring in biomathematics and electrical engineering. I currently work at the Space Dynamics Laboratory in Logan, Utah, where I support various missions relating to air, land, sea, and space. I practically live in the forest with how much I hike, and I'm notorious for trying new things, so much so that my friends like to joke that I'm the female version of the Dos Equis "Most Interesting Man in the World."

I like to think that before I was born my father knew that though I would never grow taller than my 5'5" stature, I'd take on leadership roles with the goal of changing the world; hence living up to the meaning of my Gaelic name: "little king." When you meet my family, my personality makes sense. My father is a fire-spitting Texan who has committed his entire life to being in service of others, and fighting for those who cannot fight for themselves. He's a man of pure steel and has a heart of gold. He told me that by the time I moved out at age 18, I'd have three things: an understanding of money, an understanding of how the world works, and a thick skin.

My mother reigns from the rural countryside of Rigby, Idaho. Growing up in a family of ranchers and farmhands, she knows how to stand up for herself, and has never shied away from a challenge. Despite not having an engineering or business degree, she has fought her way to becoming a program manager at Northrop Grumman. Her determination will never not be inspiring to me.

I have one older brother who doubles as my best friend. He's a bragging point whenever I discuss my family, as he is a geophysicist. We're a small four-person squad, and they are my everything.

I grew up in West Point, Utah; a quiet and rural country town that has been becoming increasingly residential over the past decade. This isn't the military West Point by the way, this is the West Point of big grass fields, loud mooing cows, and the best view of every fireworks show on the Wasatch front. I love my hometown and the people in it. Neighbors are truly neighbors, and everyone has each other's back. But people don't really leave West Point; in the very least they do not travel far. You go there to retire and die, or you stay and never leave.

I grew up in comfort and with happy core memories, but even as a child I knew I'd leave one day for good. As much as I loved the idea of staying home forever, basking in the scent of hay, watching the most beautiful sunrises and sunsets you'll ever see, my place at home always felt temporary.

I was 10 years old when my mother signed my brother and me up for summer camp. Summer camp had been an integral experience for her as a child, and she didn't want us to miss out on that experience. So off we went to Astro Camp for a week. Properly assumed by the name, Astro Camp was a camp dedicated to team building activities, all under the guise of learning about space and space exploration. The camp had been running for years and continued to do so until 2016 (and I attended right up to their last camp). It was hosted in an elementary school; easily identifiable by the giant space shuttle sculpture protruding out of the school.

On day 2 of camp, I experienced that canonical moment that everyone has in life where everything changes. Up until that point I was dead set on becoming a chef (funny now because anyone who knows me knows I cannot cook to save my life). I declared I was going to do everything in my power to be a NASA astronaut, and that a path of engineering was the path for me. It's been 12 years, and that dream has not wavered. To Ed and Lois Douglas, the camp directors: thank you. You created an inspiring environment for children to thrive and see themselves in the future of STEM (Science, Technology, Engineering, and Math). I am who I am today because of your work. From the bottom of my heart, thank you.

The importance of teamwork and kindness weren't the only core lessons I learned from attending Astro Camp for so many years. On the patch logo for the camp were the words: "Our task is not to foresee the future, but to enable it." This mindset has been vital in all aspects of my life; professionally and personally.

People tend to think I'm a morbid pessimist when they first hear my life's philosophy. That is because I think about death every single day. I let the concept of death guide me in my daily decision making. Every morning, I ask myself: "If I die tonight, tomorrow, 5 years from now, or 50 years from now, what am I going to do today to make sure

I die without any regrets?" Since worshipping this mentality, my life has transcended. I've pushed myself outside of my comfort zone, and thus been granted so many opportunities. I've had the most amazing life-changing and enhancing experiences, all thanks to being motivated by the concept of death.

I live my life in the service of others. If I can inspire one individual to see their potential of greatness, or if anybody's life has been improved by my being in it, then I have lived my life exactly as I wanted. We all want to leave behind some big romantic legacy, but I have found that making someone smile or feel comfortable in their own skin is the only legacy I need.

I decided to go into the STEM workforce purely due to the awe of space exploration and the idea of unlocking the mysteries of the universe, but what has kept me in STEM has been the notion that I can leave this Earth better than how I came into it. I have the power to change the world and encourage someone else to do the same.

Anytime I tell someone that I'm in engineering, I get the classic response of, "Oh my, so you're really smart! I could never do that, I'm not smart enough." This response has always left a sour taste in my mouth. I'm not in engineering because I'm naturally smart. The truth is I have struggled with math my entire life. Countless failed tests, letters being sent home encouraging tutoring; my high school math teacher even recommended I do not continue onto the next expected course for my senior year (P.S. I did anyway and ended up excelling).

So why go into engineering? One would think the writing would be on the wall that I was not meant to be an engineer. I chose engineering as a career because I loved problem solving and being hands-on all in the name of improving the world we live in. In all honesty, I probably struggle more than my peers in my engineering courses. I've had to retake more classes than I'd like to admit. The concepts we learn in class do not come naturally to me, but I stay in it because I love learning. Don't get me wrong, I'm not the quickest learner; but I learn. I believe everyone can achieve anything; it just takes determination and passion!

I must admit there is a hint of spite sprinkled into my motivation. I call it my "positive toxic trait." I don't give up, even when I should. Nothing motivates me more than the words "you can't do it." The word "no" is the ignition my flame needs to rage an inferno. "You can't do it" is inevitably met with my "well, I guess you're just going to have to watch me do it."

I continually push the edge until I find it. I will move forward until I break. Then I'll take a step back, reevaluate, check my resources, and go right back onto the path forward, pushing the edge once again. The human mind grows, adapts, and overcomes throughout its lifetime. We learn from our mistakes, we change our behavior, we change for the better, and become the best version of ourselves. I understand that this approach to life can be extreme and not suited for everybody. I often live my life in a very "zero to 100" manner. I'm either all in or all out.

The precursor to my college experience was graduating from the Northern Utah Academy for Math, Engineering, and Science (NUAMES). I didn't get the classic American high school experience. We didn't have any sports teams, lockers, prom queen, or even a cafeteria. Our school was inside Weber State University, and I took college classes with the big kids. I adored my high school experience. I'm still close with many of my friends from that time.

NUAMES was my first practice in getting outside my of comfort zone. I joined our FIRST Robotics team and had the time of my life doing so. I was surrounded by supportive teammates, about 45% of which were female. I learned that I was capable of doing so much more than I initially believed I could. My short time on the team opened several doors for me later in life. I graduated class of 2019 with my high school diploma and my Associates of Science from Weber State University.

I wish I could say I had some large inspirational pull towards Utah State University, but the truth is that I never considered anywhere else. I knew I'd stay in Utah, and some family had been Aggie alumni. It wasn't until a few months before moving to Logan, Utah, that I understood how great of an engineering program we have. I learned that USU was a "space land grant" university, which piqued my interest.

The Space Dynamics Laboratory is a proud university-affiliated research center, one of 14 in the country. We're sponsored by the Department of Defense, specifically the Missile Defense Agency, and have large ties with the DoD and NASA. Founded a year after NASA,

SDL has been solving unique challenges faced by the military, science community, and aerospace industry for more than six decades. Our expertise is in satellites, sensor/instrument systems, ground systems and data processing, and advanced autonomous systems. At SDL, we have a large internal research and development (IR&D) division and are proud to have several patents in the industry.

I am currently an engineering assistant at the lab, on track to be promoted to a full-time engineer upon graduation. Ideally, I'll stay with the lab as I move into a master's program. I genuinely love my job. Every day I go to work knowing that I get to be a small yet crucial cog in the overall machine that is the laboratory. I have never felt isolated here, and it is incredibly refreshing to work somewhere where I am treated as an equal, and where my ideas are heard and valued.

In my position, I design, fabricate, and test satellite components used for heat transfer applications in space. I love coming to work knowing that the hands-on work I do will one day orbit in space. Sometimes I'm reviewing electronic board documents, suiting up in a clean room gown and cleaning parts, running calibration tests on our 8-kilowatt laser, or designing and building thermal straps. I love it, and I love being able to have a hand in multiple projects.

I'm particularly grateful for the constant change of my career, because going into engineering I was terrified that it wouldn't be what I expected. I dreaded the idea that I'd spend my time glued to a desk. I subscribe to the strong belief that you were not born to pay bills and die. I'm a very active person, and every single manager I've ever had has described me as the Energizer bunny. I am not built to stay still. Sometimes we don't have control over what our day-to-day work entails, but we can control the balance of that life.

It took me about 2 years to land my job at the Space Dynamics Laboratory. I was the squeaky wheel always contacting and networking with SDL's HR team. By the time I walked into the interview for the position I have now, the interviewers were telling me: "We've heard so much about you!" I owe this to my university's chapter of the Society of Women Engineers (SWE). Upon coming to USU, I threw myself into SWE. I was in a new town, far from my family, with little to no friends there with me, and surrounded by people who had very different values and cultures from me. I was riddled in loneliness and found purpose and comfort in SWE. Not only did I find deep and meaningful friendships with my SWEsters, but I was introduced into the large industry that is engineering. My best connections have come from meeting women in engineering at SWE conferences. I truly owe almost, if not all, of my professional opportunities to SWE.

I had had several interviews for different positions at SDL and always got the "thanks but no thanks" email. Of course, it was always devastating. I was desperate to get out of the retail industry and begin my career. I have vivid memories sweeping a deli floor on my closing shifts, reeking of fryer oil, dreaming of the day I'd get out. But I held true to the idea that it's not rejection, just redirection, and kept trying.

The rejection of not getting the jobs I applied for only pushed me to keep applying for other positions. Life is all about taking the stepping stones of opportunity. I did eventually get out of retail, moving to work as an assembly worker at a local machine shop, which I adored. I fell in love with manufacturing. Every day I got to work on parts serving different purposes from the aerospace, or medical industry, to firearm components. I didn't ever expect to love manufacturing so much, but being in that type of engineering made me feel like Scotty from *Star Trek*. I have learned time and time again that the skills, hobbies, and jobs that you would not expect to enjoy sometimes end up being the greatest times in your life. It broke my heart to leave that job, though it was for the better to be a summer intern for a big aerospace composites company.

One of the greatest lessons I learned from being an intern is that no matter what job you have, it will always feel like you are drinking from a firehose. Nobody jumps into a job position fully prepared and knowing all there is to know. The big secret? Everyone is faking it and hoping they don't get caught. We're all stumbling through life, doing our best with what we are given. It was a tough lesson to learn, but a valuable one knowing that it is okay to be vulnerable. It's okay to admit when you don't know something, to ask for help, and to ask for repetition.

Being an intern at that company was tough. I was a sophomore intern among juniors and seniors who had a much more extensive understanding of engineering than me. My imposter complex was almost crippling. If I could go back in time, I'd tell younger me: "you deserve to be here just as much as the other interns, and you deserve to get as much out of this as them."

I was put on a team in charge of one of the company's biggest projects. To say I was thrown into the deep end and told to swim is an understatement. There was simply not enough time in the schedule to allocate towards training me to do the tasks I was assigned. I was stumbling through my internship, hoping eventually my feet would keep me steady. I made several mistakes. Other teams would reassure me I was doing great, but I could see the deadlines we weren't meeting, and I would spend my 9-hour days making little to no progress.

It was blatantly obvious my supervisor was frustrated with me. The project was in shambles for many reasons, but adding a seemingly incompetent intern to the mix didn't help with the pressure he was probably feeling. Unfortunately, I became his punching bag. Rather than being trained to do better, my responsibilities would be taken away from me until my role at the company boiled down to secretarial work.

The most hurtful thing among the many sexist and outright mean things my supervisor said to me over the course of my internship, was a reaction to me explaining to him how I fixed a mistake in our project. He looked me in the eyes for a few moments, not saying anything, then broke the silence with:

"I hope you never become an engineer because so many people will die."

Being the people-pleaser that I am (and am actively working to fix), I laughed it off, even though my heart sank like a rock in water. I could feel myself cracking from the pressure. I didn't know how much longer I could tell myself this is what interns deal with.

My breaking point came the day an anomaly of a mistake was made, and I had to carry the responsibility of it. I could handle the horrific comments from my supervisor, but I broke when that mistake caused trouble for my coworker. The idea of disappointing the one person who I felt was on my side was detrimental. I felt like I had a stone caught in my throat as I fought to keep my composure while being coached by him and my supervisor.

I remember sitting outside during my lunch break, desperately trying to calm myself down so I wouldn't have to go back inside with red puffy eyes and wet cheeks. Crying on my lunch breaks had grown to be a frequent event. I felt so defeated. Maybe my supervisor was right? If I couldn't even manage to program an Excel formula, how was I going to be a successful engineer? How could I possibly create the change I wanted to see in this world? It was a level of shame I had never felt before. I've always taken pride in my work ethic, and yet here I was seemingly upsetting everyone I worked with, and ruining everything I touched.

My experience taught me the importance of sticking up for myself and being my own advocate. I wasn't incompetent, I was doing the best with the resources I had. I deserved the same amount of respect as anyone else there. It taught me that people often project their insecurities onto you, and that it is not a reflection of you.

I did end up sticking up for myself. I requested a transfer onto another team. The HR staff was very gracious in granting this transfer, and I am happy to say I spent the remainder of my internship having so much fun, growing very close with my coworkers, and doing work that felt meaningful to me. If you are to learn anything from me, do not wait until the pain becomes unbearable to seek help. Suffering is not a requirement and shouldn't be an expectation to being in a STEM career.

My internship concluded, and I was scrambling to find another job. Since the time I was 16, I always had a job, so being unemployed gave me tremendous anxiety. I scrolled through our university job postings, specifically ones in undergraduate research. I did not think I was smart enough for undergraduate research, but my tuition bill was a great motivator for me putting in an application. That's how I came to be an undergraduate researcher in our Experimental Fluid Dynamics Laboratory (EFDL). I ended up getting my job at the Space Dynamics Laboratory a month later and worked at both for about a year.

Once again, I wish I had an inspirational story about what attracted me to working in the EFDL. If I'm being fully transparent, I really had no interest in fluid mechanics. It wasn't until I began the project that I fell in love with the dynamics of fluids and began to see it in the natural phenomena of the world. My project was focused on creating an effective outreach method to teach the public of all ages about fluid dynamic vortex rings. It was a fun challenge, and teaching 5-year-olds about no-slip conditions and frictional layers of fluids was even more fun.

Keeping in theme with my zero to 100, "why not" attitude, I decided to throw my hat in the ring for the Society of Women Engineer's collegiate research competition. I applied using this research project as my entry. I didn't think I stood a chance, so when the email came a month later saying I was a finalist, I stared at my computer screen just blinking for some time.

I spent a lot of time preparing my research poster and research presentation. It was something I had never done before, and it excited and terrified me. When it came time to travel to Seattle and present to industry leaders and other collegiate competitors, I was ready. My research project was in no way as technical as the other competitors, but I was simply excited to be there. I was also competing with two of my fellow SWE members, though it didn't feel a shred competitive. I didn't end up placing, but I was so thankful for the opportunity to be there, and ended up making so many valuable connections.

It was in Seattle where I met the author, Dr. Justina Sanchez, as well as the inspirational and decorated Dr. Tracy Nguyen. Through various butterfly effects, thanks to Dr. Nguyen, I fell into the position of being the local counselor for my beloved Utah SWENext organization. It's funny how you never know where you'll end up based on the people you meet or the decisions you make. The girls in Utah SWENext are some of the greatest young women you will ever meet. It is a breath of fresh air to know that they are going to be the leaders of tomorrow.

If you ever feel hopeless, or that you don't belong in STEM, go talk to a SWENext group, or meet with any local STEM club. You will see what our future looks like, and it is filled with so many determined young women with dreams brighter than the stars, and a fire within their hearts to ignite those dreams. Talking to the younger generation coming up in STEM always reminds me of why I began my journey in the first place, and keeps me going when the days get hard.

The College of Engineering at USU was built in 1903. It's easy to tell too, mainly from its old architectural design, but inside the original building tells secrets of the past. Today, the college's percentage of women is 16%, which is about 350 students. It's a joke we have, that you can tell there used to be no women in our engineering program. How can you tell? The female restrooms are in very odd places of the building. Men's and women's restrooms are not side by side like one would expect. They're on different floors, with women's restrooms being across the building. They're small, cramped, and it's obvious some of them used to be utility closets. I remember laughing about this to my father one day. He told me:

"When you shatter glass ceilings, the shards tend to fall in weird places."

He arguably said the most poetic sentence I had ever heard so nonchalantly. I love this quote from him, and I think about it all the time. We often get so caught up in our advocacy and tunnel vision of creating a better future, that we forget to express gratitude for our past and how far we've come. While there is still so much progress to be made in creating a better life for women and minorities in engineering, I am grateful for all the progress that has been made. I can walk into my engineering classes, and though I am still one of maybe 10 women in the class, no one will ask me if I'm in the wrong room, or shame me into leaving. I am standing on the shoulders of the giants who paved this path for me.

Weirdly designed restrooms and amazing ice cream aren't the only things we have at USU. We have a long running tradition of Miss USU. Miss USU is a pageant that has been going on for years. Individuals representing the colleges in the university and diversity groups compete. The crowned winner spends the consequent academic year working on her platform to improve the university and surrounding community, while acting as a representative of the student body.

I had always wanted to do a pageant in my life. My mother had done them growing up, but never thought of introducing them to me, probably to avoid becoming a *Toddlers & Tiaras* cliché; ironic because

I loved watching those TLC shows as a little girl. Depending on when and where you meet me, there's a good chance you would never know that I can be hyper-feminine. I'm an avid outdoorswoman, yet I love elegant gowns and makeup just as much as grueling sweaty hikes and sleeping on hard ground under the stars. Like most things in my life, I take the approach of "why not?" So, I applied to the pageant to run as Miss Engineering, representing my college.

To be selected as Miss Engineering was an honor. I adore my college and was so proud to represent the student body and faculty of my college. I entered with my platform of creating a more inclusive environment for women and minorities in STEM on our statewide campuses. The journey of working towards the pageant was so much fun. The women I shared the stage with were so kind, fun, and motivated individuals who radiated positivity and love.

The pageant itself was the experience of a lifetime. I had never felt so confident in my own skin and was shocked to see the number of my friends who showed up, and for several of my family members who watched the livestream at home. I couldn't fathom this amount of people supporting me. When we all lined up on the stage during the crowning, I was truly happy. I didn't think I'd win at all, but I felt like I was the biggest winner anyways simply from all of the support. I held the hands of the gals next to me, ready to scream and cheer when one of their names was said. Time slowed down after my name was called. Wait, did they seriously just say my name after the words "You're 2023-2024 Miss USU?" Did I hear that right?

I'd like to note that growing up I was not confident. It wasn't until college that I began to accept myself and feel comfortable in my own skin. As a little girl I was a cliché in the sense that I cared about what boys thought of me, and really what anyone thought about me. For the longest time I lived for other people rather than myself. Suddenly I was that little girl again. I was that little girl riddled with insecurity, convinced I was too ugly, too nerdy, and too weird to be loved. But there I was standing on that stage, looking out into a crowd all on their feet screaming and clapping. I burst into tears. I felt like an actress who won an Oscar. It's important to express that I'm not a standard pageant girl. I'm the antithesis of tan, I wear glasses with lenses so thick its humbling, my hair is a pixie cut, and I've got big and visible tattoos on my upper torso. Yet, apparently, looks didn't matter after all, and the judges heard my platform and believed in it and in me.

Overnight I went from being well known within my college, due to my involvement in SWE, to being recognized all over campus. People would stop me and ask if I was Miss USU, then proceed to tell me what my winning meant to them. My life flipped on its head in the greatest way imaginable. I found myself in this incredible position where I had the power to make genuine change in my community, and to inspire greatness in anyone I interacted with.

In the 20+ years that the pageant has been running, I am the first Miss Engineering to ever win the title. When I won, you could feel the shift in what Miss USU meant. As my peers have been quoted as saying to me, "You're not just another pretty face. You're taking this position and making it really mean something."

Wearing the crown has been one of the greatest blessings in my life. I've made it my mission to make women and minorities in STEM degrees across our statewide campuses feel welcome and included.

By the time this book is published, I hope to have done more outreach with the Diné-Navajo nation, as I am saddened that indigenous people make up only 0.4% of engineers in this country, and I'd like to change that. I've also worked hard to increase outreach efforts to middle school-aged girls, as this is commonly the time when young women are discouraged from pursuing a degree in STEM due to peer pressure, a lack of a role models, and a misrepresentation of what a STEM career entails.

I memorize every interaction I have with a woman or young girl. After all this time it's still unbelievable to hear someone tell me that I inspire them, that they look up to me, and want to be me. I don't think I'll ever get used to it, and I don't want to. It is the deepest privilege to show women of all ages that they do not have to sacrifice their femininity to be a scientist or engineer. Our society has taught us to attach our worth and beauty to our physical appearance and feminine behaviors. Nowadays it's not explicitly said, but there is still a strong narrative that if you cut your hair, live a childfree life, become a working mom, or pursue a male-dominated career, you are automatically stripped of your femininity.

Our world is obsessed with putting everything in a box and demanding an existence of living in black and white, when in reality, everything is gray. Two things can be right at the same time, and I believe I am a living example of it. You can be an engineer and wear gowns (though not on the machine floor, as that is probably an OSHA violation). You can have a fulfilling career and a family. You can be childfree and still be a worthy woman. You can have half of your head buzzcut and be a pageant queen.

Nobody can stand where they are today and say they did it completely on their own. The road to my success is paved in so much sweat, tears, and numerous sacrifices several people made for me. When I look behind me, I can see an entire audience of every person who has ever supported me. I have several mentors who have been wonderful guides throughout my professional life. So many of my opportunities have only become available to me through Dr. Elizabeth Vargis, Brooke Mckenna, Nina Glaittli, Carmen Park, Emily Andersen, Dr. Tracy Nguyen, Zac Humes, and Dr. Tim Berk; to name a few.

The support doesn't end with your professional contacts. It's not all about who can get you an interview or who says your name in a room full of opportunity. The people who see you at your lowest are equally as important as those who stick with you at your highest. What you see on Instagram, LinkedIn, or at face value is life's highlight reel. What I often don't show people is all the nights I spent sobbing so hard I couldn't breathe; The endless days I struggled with depression and suicidal ideologies; the times I failed hard even after giving it my all; or the times I had to face the heartbreak because holding on would only hold me back. I'd like to give a special shoutout to the integral people who stayed on the call as I cried, who sat with me for hours to let me rant, and who never let me lose sight of my dreams and who I was: Sydney Leneweaver, Crystal Tingle, Daniella Rivera, Mackenzie Helmick, Kevin Thomas, Cooper Brayton and my brother, Ryker Tracy.

I'd be remiss if I didn't also mention my grandfather Dwight Walker, and my dearly departed great-grandmother, Sharon Walker. My grandfather is a life-long learner. He is a huge advocate for education.

Grandpa doesn't talk about it much, but he holds a bachelor's degree in industrial technology with an emphasis in nuclear power plant operations. He taught me everything I know about fishing and botany, but he also taught me how to never stop learning. He never gets tired of hearing about my work or my dreams, and he is arguably the biggest cheerleader I have.

My great-grandmother was one hell of a woman. She was the sweetest person you'd ever meet, and always stood up for what was right. She dedicated her life to helping others, and she never knew a stranger. Towards the end of her life, she always had visitors. While her memory had been fading slowly but surely, she never forgot my dream of working for NASA. I'd walk in the door, and she'd introduce me to her guests with, "this is my granddaughter, she's going to work for NASA one day!" She made me promise not to repeat her mistakes. I promised her I would not marry until I had my bachelor's degree, I'd never prioritize a boyfriend, and that I'd live my life as much as possible. I can only hope I would've made her proud.

I will spend the rest of my life wondering what I could've possibly done to deserve all the love and support I've been given. I am incredibly lucky to have this life, and it is a privilege I do not take for granted. Sometimes I wonder if this is it, if life can get any better, and then it somehow does. I'd like to think if I keep putting good into the world, the universe will return good back into my life.

I end this chapter of my story with this: If you find that there is not a seat for you at the table, build one. You've got the brain and creativity of an engineer. You can design any seat you want and make it as large as you'd like. Do not be afraid to take up space. While you're at it, build a few more for the women who will follow in your footsteps.

This life is truly a gift. The probability of being born is 1 in 400 trillion. Your chances of existing is so infinitesimally low, yet here you are. Don't just run towards your dreams, sprint! And if anyone stands in your way, barrel through them. Life is too short and too precious to spend it doubting your capabilities. Trust your gut, and do not let fear stand in the way of leaning into all the opportunities presented to you. I promise if it seems scary, that's just more reason to do it. As Cleopatra says in Shakespeare's *Antony and Cleopatra*: "What's brave, what's noble, let's do it after the high Roman fashion, and make Death proud to take us." Make Death proud to take you. Live a life worth dreaming and talking about. You owe it to yourself to at least try.

Connect with Regan:

An Accidental Engineer

Coauthored by Jill S. Tietjen, P.E.

I AM AN ACCIDENTAL engineer. Although my father was an engineer, as the oldest of four children and getting ready to go to college in 1972, no one – no one in my family, no one at my school, no one in my community – suggested that with my interests and aptitudes that I should consider getting an engineering education. Fortunately, I did eventually find engineering, which was the right educational and career choice for me. But I am getting ahead of myself. A little background would be helpful.

I grew up in Hampton, Virginia. I started working on jigsaw puzzles when I was two years old. By age three, I must have been a little precocious. My mother asked around regarding preschools. She was told the best one was the laboratory school (the place where the teachers were college students doing their student teaching) at Hampton Institute (today Hampton University). In 1958, right after I turned three, she enrolled me. I attended three-year-olds, four-year-olds, and kindergarten there and got a fabulous foundation for my learning and education. I like to tell people that I am an alum of Hampton Institute!

I became a Brownie at 7 years old (this was before there were Daisies), and stayed in Girl Scouts until my troop dissolved when I was 14 years old. Today, I am a lifetime Girl Scout and have served as chair of the board of directors of the Girl Scout, council in Denver, Colorado. At eight years old, I started playing the violin. All through elementary school, junior high school, and high school, I played in the community orchestra – there was no school orchestra. Later, I played in the University of Virginia orchestra and in community orchestras in North Carolina and Colorado. I played for 33 years in total. I was very involved in extracurricular activities, often in a leadership position. There were no girls' sports teams for me to participate in during my junior high and high school years, as this was before 1972 and the passage of Title IX.

My siblings and I were raised to do our *personal* best – not to be *the* best. As we were growing up in Virginia and my father was a federal government employee, the four of us were told we were going to college and that we would be going in-state. My parents planned to pay for most of our college expenses and they planned to have enough money to pay for all four of us.

In high school, at age 16, I was teaching myself FORTRAN with self-guided workbooks that my father got for me. I was also a good writer and liked writing – I scored well on Advanced Placement tests in both calculus and English. I entered the College of Arts and Sciences at the University of Virginia as a math major because I loved calculus. I don't really care for the sciences – and I especially don't like biology and chemistry. But my first semester in college I discovered that I was in the wrong place – I needed to be in the School of Engineering and Applied Science. I wanted to apply my math knowledge and solve worldly problems – not pursue a career in research and teaching. I made all the necessary arrangements with the dean of the School of Engineering and Applied Science, and transferred in at the beginning of my sophomore year. I loved engineering and attending the University of Virginia (I still sit on the engineering school's Board of Trustees!).

I graduated with a degree in Applied Mathematics and a minor in Electrical Engineering, and went to work in the System Planning Department at Duke Power Company in Charlotte, North Carolina. It was a perfect fit for me. I became a utility planner – a title that I still proudly wear more than 47 years later! I worked with simulation models – models that captured the manner in which the power plants in the Duke Power fleet operated at that time, and with appropriate forecasting data, into the future. I worked with production costing models – modeling day-to-day operations in hourly increments for now and the next few years – and capacity expansion models – models that simulated the operation of the system for twenty years, and provided options for when new power plants should be installed, what type of fuel they should have, and how big they should be. I first studied renewable resources in 1979 – including solar, wind, tidal power, ocean thermal, geothermal, superconductors, batteries, and many other technologies.

I got married right after I graduated from college. My mother-in-law and father-in-law died about nine months after we got married and when we were both 22. We finished raising my first husband's two brothers who were 14 and 18. I also had my first experience with a cat as a pet – I grew up with dogs.

My bosses recognized that I had the ability to speak and that I enjoyed it. They saw to it that I was trained as a speaker, and I served as a member of Duke Power Company's Speaker's Bureau – going out and speaking to groups around the service territory about energy conservation, nuclear power, and related topics. I was also selected as a member of the crisis management team – the team that assists during simulated emergencies at nuclear power plants. My role was a technical briefer – which meant I explained to the journalists in plain English what the engineers had just said, so the reporters could write in the newspaper or report on television about the event. During my years at Duke Power, I took advantage of their tuition reimbursement program and earned my Master of Business Administration at night at the University of North Carolina at Charlotte. That degree has proven invaluable as a necessary credential for my subsequent jobs and board positions.

I discovered the Society of Women Engineers (SWE) at a career fair at North Carolina State University (where I was doing on-campus recruiting for Duke Power) in 1979 while I was living in Charlotte, North Carolina and worked to establish the Charlotte-Metrolina Section. SWE's primary activities – career guidance and professional development – dovetailed with my new determination that women needed to be encouraged to be engineers (as I had not), and that once they were in the field, they needed opportunities to develop many skills in a safe, supportive environment.

I moved to Denver, Colorado in 1981 and went to work for Mobil Oil Corporation's Mining and Coal Division. I learned the coal business upside down and backwards in this job. I also became very active in SWE in the Denver area, and I became licensed as a professional engineer in Colorado. Then in 1984, I began my consulting career. Also, in 1984 or 1985, my husband and I tried to start a family of our own. We were not successful even after many infertility tests and treatments.

Around 1987, I started serving as an expert witness in hearings before regulatory commissions around the country. I benefited tremendously from witness preparation, especially the sixteen hours I spent getting ready to be on the stand in Maine. At the end of that weekend, I could answer any question (and can still today) you asked me what the answer was (the answer I wanted to provide), and complete the answer with "so that we can provide you with safe, economic and reliable power." I also wrote a number of technical papers, gave conference presentations, and honed both my speaking and writing skills.

In 1987, my SWE colleague and good friend, petroleum engineer Alexis Swoboda came back to Denver from the SWE national convention in Kansas City, Missouri, and told me she had heard about an outreach program that she wanted us to implement in Colorado. The idea was an essay contest on Great Women in Engineering and Science. It was a wonderful idea, but I didn't know who they were. Thus began our research and writing about historical and contemporary women in STEM (Science, Technology, Engineering, and Math). We wrote articles for magazines and newsletters, we gave conference presentations, we wrote nominations, and we developed the foundation for books that I would later write. We also documented the entire process of establishing the essay contest as an outreach or career guidance program so that others could implement the program themselves without having to reinvent the wheel.

My partnership with Alexis also resulted in a scholarship program for the Rocky Mountain Section of SWE. We collaborated with others to raise money to endow scholarships that, once established, were named for pioneering women in the section. In addition, I decided to establish a scholarship in my name at the national level while I was still alive. What fun it has been to meet and become friendly with some of the recipients of my scholarship!

SWE has provided me with opportunities to develop skills that I have used throughout my career in my jobs, in the organizations with which I was affiliated, and for the boards on which I have sat. These skills include strategic planning, facilitation, budgeting, managing and motivating colleagues, scheduling, and many others. Co-chairing and serving as Treasurer of SWE's 2001 National Convention held in Denver solidified and strengthened all those skills.

I am proud of my SWE legacy, especially the trajectory of Heather Doty. Heather wrote an essay for the SWE Rocky Mountain Section Great Women in Engineering and Science essay contest when she was in sixth grade. When she was in high school, she received a certificate of merit from the Rocky Mountain Section of SWE recognizing her academic achievements in math and science. She attended college at the University of Colorado at Boulder – where she served as my paid student assistant when I was the Director of the Women in Engineering Program at CU-Boulder. She is also one of the people profiled in my introduction to engineering book, *Keys to Engineering Success*. Heather became very active in the Rocky Mountain Section of SWE when she started in her professional career, and was subsequently elected to the Society's board of directors. In 2020-2021, she served as President of SWE, nationally and internationally. She led the Society through the pandemic and ensured the organization's survival. And, maybe, just maybe, I can take a little credit for influencing her and/or serving as a role model for her along the way.

In addition to Heather, I have mentored many other women over the years – through SWE, through MentorNet, through BoardBound, and informally as well. It gives me great satisfaction to help provide advice and direction and, many times, just to sit and listen and serve as a sounding board.

In 1988, I was elected to the national board of directors of the SWE. Through that organization, I began nominating women for national awards using the knowledge I had gained from researching women for the essay contest. My first big success was nominating Admiral Grace Murray Hopper for the National Medal of Technology – the U.S. equivalent of the Nobel Prize, awarded by the President of the United States. Admiral Hopper was in ill health and asked me to receive the medal on her behalf – which I did at the White House Rose Garden in 1991. As I looked around at the award recipients, I wondered where the women were – Admiral Hopper was the first individual woman to receive the National Medal of Technology. I made it my personal mission to submit nominations of women for local, state, university, community, and national awards.

I served as national president of SWE during the 1991-1992 fiscal year. My first successful nomination to the National Women's Hall of Fame was in 1994 – also Admiral Hopper. The family asked me to receive the posthumous honor and thus began my association with the National Women's Hall of Fame. I served on the National Women's

Hall of Fame board of directors 2009-2014 and as CEO of the organization in 2015.

In 1994, I got divorced from my first husband and met the man who became my second husband. In 2023, after 27 years of marriage, a divorce again is in process. I expect the divorce to be finalized in 2024.

I started being asked to serve on university and other boards because of my SWE board participation and having served as national president. One outcome was that I was recommended for and in 1997 was elected to the board of directors of Georgia Transmission Corporation, an electric utility in Georgia. After serving for several years on the external advisory board to the Women in Engineering Program at CU-Boulder, I applied for and was selected as its director. I served in that position from mid-1997 through 2000. I wanted to know if my passion for women in engineering should be my career focus – but decided that it was not and that I still wanted to do my engineering consulting.

When I left that CU-Boulder position effective the beginning of 2001, I started my own company. Also in 2001, my first books of which I was a co-author were published. One book was an introduction to engineering book titled, *Keys to Engineering Success*. The others were in the Setting the Record Straight series. The first book in that series focused on the women's rights movement and women's entry into non-traditional professions. The second was a history of women in engineering. I say that although I was unable to have children, I give birth to books – 15 as of mid-2024, with more in process.

Now my engineering career and my writing career are functioning in parallel. I was still advising clients, serving as an expert witness, and overseeing the results of production costing and capacity expansion modeling. Numerous power plants are in operation around the West and Midwest that I had a role in permitting and getting into the company's rate base. In parallel, I began to write more books, speak, and write columns for *The Huffington Post*. The focus of my writing became writing women into history, once I realized that women had been omitted from almost all written history in the U.S. and around the world. I considered that a serious oversight that needed to be corrected and that I could do it. Interestingly, I use my engineering know-how in putting together spreadsheets and keeping track of all of the information required to write these non-fiction books. I say that I solve book puzzles – which are very similar to word problems throughout math and engineering. Which parts of the puzzle pieces are relevant, how do they fit together, and how do I make a complete book?

In 2014, out of the blue, I received an email from an editor at Springer, an international publisher. She had seen an article that I had written for an encyclopedia published by Taylor and Francis, and wanted to know if I was interested in writing a book for Springer on distributed energy. Would I be willing to have a conversation with her? We did have a conversation and the outcome was that I was not going to write a book on distributed energy. What I was going to do was update and revise the history of women in engineering book that I had written for the Setting the Record Straight series, and Springer would publish the revision. It would serve as the foundational volume for the Springer Women in Engineering and Science series, and I would serve as the series editor. It would be my job to recruit volume editors. These volume editors, who would all be women, could pick any topic within the broad umbrella of engineering and science. They would each recruit a minimum of eight to ten other women who would write chapters on the topic the volume editor had selected. As of late 2023, around 25 books have been published in the series, a number are under contract or development, and other potential volume editors are in the pipeline.

This is one of my examples throughout my life of seizing opportunities. In fact, my ebook in the Institute of Electrical and Electronics Engineers (IEEE) women in engineering series, Book #9, is titled *Recognizing and Taking Advantage of Opportunities*. Before you can take advantage of opportunities, you must have your eyes, your heart, and your mind wide open so that you can recognize an opportunity has presented itself. Or you might not recognize it for the opportunity that it is but you still say "yes." And remember what Thomas Edison said, "Opportunity is missed by most people because it is dressed in overalls and looks like work."

I didn't recognize the essay contest or preparing nominations as opportunities, but they were. If I hadn't put the work in, I wouldn't have been able to capitalize on them or build from them. I did put the work in. I did my very best. I under promised and overdelivered.

Today, I rarely do consulting, but I use the engineering knowledge that I gained throughout my career for my service as an outside director on the Board of Directors of Georgia Transmission Corporation. I also served on the board of an engineering firm, Merrick & Company, from 2010-2021, and used my career knowledge there as well.

I have received many honors and accolades for all of the work I have done – those opportunities that did come dressed in overalls. Included among them are Distinguished Alumni awards from the University of Virginia, the University of North Carolina at Charlotte, the University of Colorado at Boulder (in the special category since I am not an alumna), and Tau Beta Pi. I have also received the Distinguished Service Award or Distinguished Leadership Award from the Rocky Mountain Electrical League, SWE, and the National Council of Examiners for Engineering and Survey. In addition, I have been inducted into the Colorado Women's Hall of Fame, the Colorado Authors' Hall of Fame, and the National Academy of Construction.

I believe that it is incumbent on me to use the talents and abilities that I was given to their highest and best use. If I do not, I am not doing what I was put on this earth to do. I am not special – my talents and abilities are mine – make sure you use your own talents and abilities to their highest and best use as well. I may have been an accidental engineer, but today I know my mission, purpose, and direction and live and work accordingly.

Connect with Jill:

Lift As You Rise

Coauthored by Danielle Schroeder, PE, ENV SP
Professional Engineer and STEM Content Creator

Hi there, my name is Dani, and I am a civil engineer. My engineering origin story started at home - literally. My dad is a carpenter, and he built the house I grew up in. As I got older, I learned more about the construction side of things through side projects he would complete around our house, as well as hearing about the projects he was working on at his job. So, through my dad, I learned about how awesome construction could be! When I was in middle school, my favorite subject was math. I also really liked mystery novels like, *Cam Jansen* and *Encyclopedia Brown.*

As I grew older and went into high school, I still really liked math and learned to like science more. I went to a small all-girls high school in the suburbs outside of Philadelphia. One of my favorite classes in high school was my Honors Physics class. One day, we had an outdoor class to push a car while we were learning about acceleration. We graphed the change in position over time and using integrals and calculated the acceleration, which was a cool way to reinforce what we would learn in the classroom, by applying it to the real world. I loved all the hands-on labs we got to do throughout this class. While I loved many of the classes I took in high school, I also had other interests including Drama Club, where I was in all the Fall Plays and Spring Musicals, as well as being on my school's tennis team.

With my working knowledge of construction from a very young age, you would have thought that I would have learned about "Engineering" as a possible career by now, but it wasn't until my junior year of high school that I would learn about engineering. That year, I attended an Engineering Girls Camp at a local college where I met young women currently studying engineering, along with professionals in a variety of engineering disciplines. I got to do a bunch of different labs and I primarily loved the ones that focused on civil engineering. This camp taught me that civil engineering would be the perfect profession for me, as it combines construction with the application of math and science through problem solving! So civil engineering was the major I chose for submitting college applications.

I applied to several colleges and ultimately chose to attend Drexel University in Philadelphia, Pennsylvania. Because Drexel incorporated co-op into my class schedule, it was a five-year program. Throughout those five years, I completed three different internships that were 6 months each so that I could find out how the classroom relates to the real world. I learned through classes and co-ops that Civil Engineering in itself is broad, and that there are so many different career paths just within Civil Engineering.

I also completed research my first year of college. For my research, I studied the stomatal conductance, or essentially the health of the plant of two species of plants, to see which species would do better in a typical, drought, and flood setting. While this was an awesome experience, I realized that research was not something I was interested in, which is why I went straight into industry after graduation. I did still pursue some higher education, as during my third of five years I applied to Drexel's BS/MS accelerated degree program and was accepted.

In my last year of college, I took the FE or the Fundamentals of Engineering exam, which is the first of two tests that many civil engineers take if they want to become a Professional Engineer.

Lastly, I was highly involved in student organizations - primarily SWE (Society of Women Engineers), and held several leadership positions during my time including Section President, while also balancing working part-time at one of my former co-ops. In January of my senior year, I interviewed at a few places, including at one of my past co-ops and ultimately joined a local private consulting firm as a bridge engineer. I graduated from Drexel University in 2017 with both my Bachelor and Master of Science in Civil Engineering.

College was not all rainbows and butterflies. While I ended up accomplishing some great things during college, I struggled with my transition from high school to college. The pace of my college classes was quicker than I expected, and I ended up withdrawing from one of my physics classes in my first year. I did end up taking the class again the following year and made sure I asked for help when I was struggling, went to office hours, organized a study group, and thankfully, passed this class the second time around.

There are also a lot of challenges beyond the technical side being a woman in STEM (Science, Technology, Engineering, and Math). The path to becoming an engineer as a woman will sometimes be lonely, which was a stark contrast to my all-girls high school. I, to this day, remember the feeling during my first year of college when I looked around the room of my computer programming class of 30 students and realized that I was the only woman. It is tough not to feel like you don't belong when you are the 'only' in a group and, unfortunately, my story is similar to many others who are currently underrepresented in STEM.

According to the U.S. Bureau of Labor Statistics, only 17 percent of civil engineers are women. The way I have persevered despite this, is by finding my support squad. My squad in college was filled with awesome fellow women engineers in all different majors that I met through SWE, who would hype me up and keep me motivated to continue with my classes.

Onto my life as an early career professional. For the first few years of my career, I worked at a private consulting firm at their headquarters in Philadelphia. One thing I didn't expect from the transition from college to the professional world is how much learning I would do on the job. My time at Drexel was great for teaching me critical thinking skills, and gave me a great foundation in the engineering design process. I learned how to break down a problem, the basics of concrete and steel design, and how to use the steel manual. These are all a part of the ABET curriculum for Engineering Departments in the U.S.

However, there is still much more you will need to learn on the job about the specific role that you were hired for. For example, one of the main clients that I work with is PennDOT. For the first few years, I learned a lot about PennDOT standards and publications, and how to apply them to the projects that I'm working on. Generally speaking, the work I completed during the first few years of my career can be divided into bridge design and bridge inspection. Bridge design is all the calculations and drawings that go into being able to bring a bridge from blueprints to being built.

One of the first main projects I worked on was the I-95 CAP or Central Access Philadelphia. This project's goal is to cap a major highway with a 12-acre park, and has just recently started construction. I have also worked on some other amazing projects including the retrofit of the Burlington Bristol bridge and the I-676 multi-bridge replacement. I also started my career doing about 5 percent bridge inspection. The

NBIS or National Bridge Inspection Standards require safety inspections at least once every 24 months for highway bridges that exceed 20 feet in total length, located on public roads. Many bridges are inspected more frequently, especially as the bridge gets older. I loved my bridge inspection work because it gave me a more holistic view of what I was designing in my other projects.

Now that you know about my engineering journey thus far, let's talk more broadly about Engineering, or the E in STEM! The broad definition I use for Engineering is that we use science and math to make things efficient, economical (or cost efficient), and safe, because safety is so important to any engineering field! Engineers are solving problems in so many fields and helping to create and improve products like prosthetic limbs, cars, airplanes, computers, and so much more. There are many types of engineering, with some of the oldest engineering disciplines including Civil Engineering, Chemical Engineering, Electrical Engineering, and Mechanical Engineering, among many others. As technologies have evolved, new engineering disciplines have been created over time, like Aerospace Engineering and Computer Engineering.

Within Civil Engineering specifically, we focus on the design, construction, and maintenance of our infrastructure. Infrastructure includes more direct things like roads, bridges, and buildings, but it also includes things like making sure you have access to clean drinking water when you turn on your sink, or have water to take a bath or when you flush the toilet. And civil engineering isn't just fancy buildings and bridges – it is also things like the foundations of buildings, like your house, school, and more. It also includes the pipes that deliver your drinking water and transport your waste.

In terms of certifications or all those letters behind my name - I applied and became an EIT or Engineer in Training after I graduated, because I had already passed the FE Exam. I also attended a two-and-a-half week-long training and passed the associated exam to become a CBSI, or certified bridge safety inspector, though I have let this certification expire since my current work no longer includes bridge inspection.

After four years of progressive engineering experience, I was then eligible to sit for the PE Exam or the Principles and Practice of Engineering exam. In 2023, I earned my PE or Professional Engineer license - though it did take me multiple attempts to pass the PE Exam. Once you earn your PE, you are no longer an Engineer in Training, which is why only PE is after my name now. I am also passionate about sustainability, which is why in 2020 I earned my ENV SP, or Envision Sustainability Professional certification.

Similar to when I was in college, I am still highly involved with SWE, as well as other professional societies. I recently completed my three-year term on the SWE Senate, which is the strategic body of SWE. I especially loved this society-level role as I got to work with awesome SWE members from around the world, who are all doing amazing things in their engineering careers.

Bringing us up to the present, I switched jobs in late 2021. While I still work on bridge retrofit and reconstruction projects, I have shifted to working on more parts of the project than just the bridge.

Outside of work, I have many passions. I love playing tennis and I have been snowboarding since 2016. Currently, my partner and I are planning our wedding which is scheduled for late 2024.

I have learned so much throughout my career thus far and love sharing what I have learned with the next generation, and encourage everyone as they progress in their career to - Lift As You Rise. As the first engineer in my family, I had to learn a lot about the engineering field in college and beyond by asking a lot of questions and seeking out mentors who are already in engineering.

I lift as I rise primarily through my STEM outreach. Since 2019, I have impacted over 7,000 students with my STEM Outreach efforts through both virtual and in-person events. I have also virtually spoken to students in 12 different countries and 14 states. One of my favorite STEM outreach programs is Transportation YOU, which is a mentoring program for girls ages 13 to 18 in the Greater Philadelphia area, who are interested in the transportation industry.

Because of my STEM Outreach initiatives, I was selected to work with Reinvented Magazine, the nation's first print magazine about Women in STEM. I, along with 11 others in STEM were selected to create the Princess with Powertools 2022 Calendar. This calendar shows girls the multifaceted reality that you can be both feminine and technically skilled, and has been distributed to students in all 50 states and 20+ countries. Several of my fellow princesses were part of Volume 1 of this book series, including Caeley Looney, founder of Reinvented Magazine.

For those reading this, I hope you are now thinking - how can I lift as I rise? How can I help the next generation get to where I am now? The answer can take many forms and does not need to include STEM Outreach if that is not something you are interested in. For example, reaching out to your graduate engineers in your group as they are setting goals and preparing for their first annual performance review can go a long way, and is just one way you can help those right behind you.

For those soon to be graduating from high school, you can help a younger student prepare for their college applications. Another way is to help a mentee with one of their goals, or even just reach out to someone at your school or company that you may be able to help along their journey! Anyone can lift as they rise by sharing what they have learned with the folks around them.

Thank you for letting me share my journey with you, and thank you Dr. J. A. Sanchez for putting together this wonderful book series. Please feel free to reach out to me with any questions you have about civil engineering.

Connect with Danielle:

Simulating Success

Coauthored by Jennifer Schmidt
Plastics Engineer at American Injection Molding (AIM)

I'm Jennifer Schmidt, and I am a Plastics Engineer. I am married and my husband is also a Plastics Engineer.... so, there's no hope for our two children, who are always steered towards STEM (Science, Technology, Engineering, and Math) activities! My family is currently settled near Erie, Pennsylvania. I grew up in a tiny town, Warren, Pennsylvania, in the Allegheny National Forest. I'm the youngest of four kids, and the only girl. I like to joke that growing up with three older brothers really prepared me for this male-dominated field of engineering. I have a Bachelor's Degree in Plastics Engineering Technology from Penn State Behrend.

I owe my engineering career to my parents. I excelled in math and science in high school and my parents basically said, "If we're going to help you financially with college, pick your favorite kind of engineering." I think they would have been a bit more flexible in reality, but I was thankful for the guidance. How the heck does one pick what they want to do for the rest of their lives at the wise age of sixteen? My parents also wanted me to have a career that I could support myself with, and not be dependent on anyone else.

I started out in chemical engineering, but in my first semester I was only taking general engineering courses (boring!). My freshman roommate was in the Plastics Engineering Technology program, and was constantly bringing home interesting plastic parts and explaining the basics to me. The fact that only weeks into our first semester she was already doing hands-on activities really piqued my interest in this field. The Penn State PLET program also offered 100% job placement and is one of only a few universities to offer this unique degree. I switched majors a month into my freshman year and have no regrets.

To get into what I do now and how I got here, I do have to go back to the beginning. My first job out of college was a hot runner company up in Vermont. I initially was hired in as a designer, but eventually moved to Project/Application Engineer (think project management). The Flow Simulation group was looking for another person, and Barb, the woman that headed up the group knew I had a Plastics Degree. Most of our colleagues were Mechanical Engineers, and nothing against Mechanical Engineers (our field is filled with them), but in my biased opinion, a good analyst really needs to understand the fundamentals of plastic materials, mold design, part design, and processing in a way most without formal education won't. She really pursued me and convinced me to interview for the role. That was nearly twenty years ago, and I've been involved with this simulation software ever since. I really owe Barb a debt of gratitude as well, because I don't think I'd be where I am right now without her initial intervention and constant mentorship through our time working together.

Currently, I am the Senior Moldflow Instructor at the American Injection Molding (AIM) Institute. My main job consists of me teaching classes on how to run Moldflow software. Autodesk® Moldflow is a software that simulates the flow of plastic and the injection molding process. It is a powerful tool that helps analysts make informed decisions about the part and mold design. It's becoming widely used in the plastics industry, but is quite a complicated software to run, and therefore requires some dedicated training.

I actually was hired on at the parent company of AIM, Beaumont Technologies, to run the software as a full-time analyst. After I was hired, I was also asked to teach these classes. In the beginning, I was probably only teaching 25% of my time, but it had slowly transitioned to more and more teaching, until that became my full-time job. If you want to learn this software, chances are you'll be coming to me.

Beyond teaching new analysts how to run the software, I also teach classes to non-analysts - think more of project engineers that are given these simulation reports and they have to read and understand them. I'm also expanding and getting more involved in the other training offerings we offer at AIM, which would be more on the fundamentals of plastics. Specifically, mold design and part design. AIM is designed really to offer training to people in the plastics industry that lack that formal degree in plastics. Remember how I said most of our industry is Mechanical Engineers or other degrees that just kind of landed here? That's our target audience. We give them some formal plastics background, which frankly is difficult to find outside of a university.

So, how does one become a Senior Moldflow Instructor? Again, I just kind of landed in this role because of an internal need within the company. It wasn't something that I set out to do. It was similar to how I landed in the simulation role to begin with. The best things in my career have honestly just fallen into my lap, and I'm eternally grateful for these opportunities. At the time when I started teaching, about ten years ago, I had about eight years of experience with software. I don't think I would have been given this opportunity if I didn't have that wealth of knowledge already. I also have my Expert Level Certification in Moldflow Insight, which at the time, less than fifty people in the world had earned. Even now, that number is barely over 100, so it's still quite a rare achievement.

I was very nervous in the beginning, having only done the customer consulting work, and never teaching the material. I found myself in that first year pouring over the material, reading all the Moldflow Theory and Concepts books, and researching anything I didn't have a firm grasp on. I think that drive and the desire to not be caught without an answer to a question from students is really what helps make me successful at this. Obviously, I still will get caught not knowing some questions posed, as students sometimes pose some very difficult but interesting scenarios, but honestly, that's what makes this fun. Anyone who claims they know everything about everything is lying. One, it's quite impossible to be an expert in every aspect of this field, and two, there's always new technology, techniques, and new information to absorb.... and there's always room to grow!

My responsibilities? I like to joke that they pay me just to talk about a subject I can talk about (and do!) endlessly! I literally get paid just to talk about a topic I very much enjoy talking about, while getting free AIM branded clothing and free lunches with the students. I legitimately fell into quite the cushy gig, all while meeting new people, and getting to hear about what interesting things they do in the plastics industry. To be fair though, I have this awesome job due to my experience and demeanor. I obviously had to have a deep understanding of the software, but also be skilled enough to explain it on a basic level. I'd argue that you have to be somewhat outgoing to engage potentially introverted students and get them involved in the learning environment. You also have to have patience, as students aren't going to get all the concepts the first time, and you end up repeating yourself a lot, or having to explain a concept in multiple ways. This is where my children have helped my career..... making me repeat myself constantly and learning patience! LOL!

Beyond the actual teaching of the classes, the behind-the-scenes responsibilities include preparation for the class and the development of the teaching materials. Until you do it, you cannot appreciate how long it takes to make a handful of PowerPoint slides. The amount of research that can go into a few minute discussion is sometimes staggering. On my off time, I'm often getting caught up on emails, perhaps discussing training options with potential students, performing research on various topics, or mentoring past and current students. Another notable piece of my job is presenting at technical conferences. So, time is spent coming up with a topic and creating content to present to different audiences at these tradeshows and conferences.

In my personal life, one of my favorite activities is to travel. I've been to quite a few countries abroad and am currently trying to get my kids to all fifty states. I've been to forty-five states and sixteen national parks, and my kids have been to forty states and fourteen national parks. In this job, I'm lucky enough to get to travel. Often times, if a company has enough people to train, it's cheaper for me to go to them, rather than have them come to me. My favorite onsite training was Google.... Just to see that campus and work there for a few days. I also got to travel to Paris for a conference and was asked to speak at a conference in Germany before it had to be cancelled due to the pandemic. Fingers crossed that I get that opportunity again. While I find myself in Michigan and the Detroit area a lot, I've also made multiple trips to California, Atlanta, Wisconsin, Connecticut, etc. When traveling for work, I enjoy spending my free evenings exploring the area, going on small hikes, and trying the local restaurants and breweries. I think it's a neat way to incorporate my personal interests with the benefits of having my work pay for my travel!

Another highlight is that I was recently asked to write a series of articles for a well-known publication in our industry- Plastics Technology. An editor from the magazine had attended one of my trade show presentations and thought it would be a great topic for their readers. I wrote a three-part series of articles that were published in multiple issues over the last year. The third piece, which they titled "Simulating Success," was chosen to be the cover story for the September issue. I was honored to be asked to write this in the first place, but ecstatic that I nabbed the cover.

As for challenges, I think my initial challenges in the role was my forementioned need to know the answers to most questions, generally not to feel stupid, and to have the validation that I did know the material enough to be considered an expert, or rather someone who was qualified to teach.

The second biggest challenge for me was my fear of speaking in public. The classroom setting never bothered me, as it's focused on smaller class sizes (typically no more than eight people), and so it is more casual. Speaking in front of large groups, sometimes a few hundred people, is what I had a hard time getting accustomed to. During my first speech I was so nervous, and my hands were shaking, so I held onto the podium to hide that fact! I think there were a few things that helped me get over this fear: one, just doing it enough that you basically desensitize yourself to it. And two, it's really a mental thing… and my own perception of that. My brother-in-law had given me good advice on the topic. He said "you're in that position because you have some knowledge to share and you really do know more than most, if not all, of your audience." I think, again, gaining that confidence that yes, I do know this topic better than most was the key for me to feel more at ease.

I'm at the point now where I am still very prepared for the given topic, but not to the point of memorization. That way, it feels like a conversation with the audience, and I can "roll with the flow" instead of freezing up if I happen to miss a beat or forget a word. Also, aren't we always our own worst critics? Chances are that small things that I will worry over likely weren't even picked up on by the audience. A few deep breaths prior and a friendly face to focus on in the audience also help! I'm certain I didn't feel it at the time, but I'm grateful to have been pushed beyond my comfort zone, as not only did it help my development, but also expanded my opportunities in my field.

Things I wish I knew before I chose this career, good or bad? I honestly wish I had a better idea of what engineering jobs could be. Again, I'm very lucky and I absolutely love what I do, but it's not like

I knew my current role was a potential career path. I think that's what makes plastics so interesting - there's so many different possibilities out there for someone with this degree. I literally talk for a living. Prior to that, I got to play with Moldflow software all day, solving puzzles for customers..... or at least that's what it felt like. It was a pretty cool feeling to have a happy customer because you helped them solve a problem.

Don't like sitting behind a desk? Others with my degree work out on the molding machines on the manufacturing floor. Some actually design products that you find in your house or cars. I feel like if I was going to get bored in my current path, I probably already would have been.... It's been eighteen years. But, if I do get bored, it's nice to know that there are a lot of other avenues to explore. Co-ops and internships are much more prevalent now than when I went to school, and I would advise students to use these as opportunities to explore multiple options so you can really understand what's out there.

As far as the bad, I find there's still sexism in this male-dominated field. I don't think I went into engineering not expecting that, but I have a few stories in my career that still caught me off guard. An old college friend was upset because I got a job offer from a company that we both interviewed with, and he did not, so he called me a diversity hire. He claimed they only hired me to meet a quota because obviously, a female could never beat him. I had known him for years, and up to that point, never seen anything that would have alluded to this kind of behavior.

A male customer refused to deal with me simply because I was a woman. My male colleague told him to shove it, and that he wouldn't be helping him (an ally!). On the flip side from that, I had a different customer that loved to seek me out for his projects, because in his words "women just think about things differently and approach problems from a different angle." I have had managers be nice to my face, but make glass ceiling jokes behind my back. I have constantly been questioned about my qualifications, or worse, shared an idea that goes ignored until a man in the room repeats it as his own.

I'm certainly not trying to steer anyone away from this field, and I'm very much of the mindset that women can accomplish anything they want. But I also think it helps knowing that it's still out there, and be prepared for how you might deal with these situations. I find that the older I get, the less I care about how people perceive me, whether it is "bossy" or "mean," which frankly can be code words for "assertive" and "holding someone accountable." I'm less likely to let the microaggressions slide. Because I work for a smaller company, and there's no immediate advancement opportunities for me, I don't need to feel like I should bite my tongue as much as I would have in my younger years. If a male colleague constantly interrupts or talks over me, I'm more likely to "reclaim my time."

If I was still working for a larger corporation, there might be more politics to navigate through in order to advance my career. I will say, I rarely have these encounters in my current job. Most of those examples are earlier in my career and were at large corporations. I'd like to think I don't see it much anymore because things are improving, and hopefully they are, but I also think it comes down to both the size of the company and the culture of the company, at least from my own personal experience.

I frankly would love to get more women in engineering. I'm involved in our local chapter of Society of Plastics Engineers (SPE), sitting on the board of directors, currently as the Treasurer. I love to get kids, girls especially, exposed to STEM and Plastics. I've helped get the PlastiVan to my town's middle school to educate the kids on what makes this field cool! I volunteer for Girl Scouts, and am one of the leaders for my daughter's troop. Although we cover many topics, I personally love sharing some of the STEM ones with the girls. I'm actually looking to get more involved at our school level, to not only share what this field entails, and that plastics aren't all evil, but also because representation matters. I believe that just from girls seeing women in this field, subconsciously or not, they are more likely to consider engineering as a path.

When I chose Plastics as a college major, I'm not sure I had much of an understanding of what potential jobs might entail. I'm quite certain I never foresaw my actual career path, nor would I have actively pursued it with my fear of public speaking. It's truly funny how life works out and how you end up where you end up! I feel lucky to have fallen into this cool little niche job and work with the wonderful people here. I'm still not sure my family and friends outside my industry really grasp what I do for a living, but when you can say "I train companies like General Motors, Space X, and Google," it sure does sound impressive!

Connect with Jennifer:

The Journey of Continuous Improvement

Coauthored by Grethel Cristhina Cabuto Sotelo

One time I read this phrase, "ultimately it's the simple things that make a difference," and that stuck with me for a very long time. Many times what we think is simply ordinary can lead to extraordinary moments that make a difference.

My name is Grethel Cabuto. I'm just a Mexican girl who is an engineer. Since I was a little kid, I dreamed about becoming an engineer. At that time, I didn't know which type of engineering I wanted to pursue, but I was 100% committed to achieving my dream. I was born in a small town called Caborca in Sonora, Mexico, but my family moved to the border city of Mexicali, Baja California, Mexico when I was just two years old. I grew up with my grandparents and my mom. They moved to look for better opportunities for the whole family.

Becoming an engineer has been the best decision in my life. It was not an easy path, but it has been a very fulfilling one. My journey as an industrial engineer led me to multiple roles across a company, including leadership roles. This gave me the opportunity to interact with people across all levels within a factory, and that's been an amazing experience. Being able to improve processes and help others is what makes me happy, and having worked with so many departments really helped me understand other people's needs.

I have been able to work in many types of industries: automotive, medical devices, aerospace defense, and now biotech. I never imagined how diverse it is to be an industrial engineer! I am a person who perseveres, and that is something that has been key to my career. I started as a manufacturing engineer, and I ended up in leadership roles after many, many hours of hard work. I have been lucky because I have been able to work in two countries, Mexico and the U.S.

I believe that we can inspire other generations just by sharing our stories. I have been mentoring future Engineers and supporting many interns during my journey as an engineer. "Be the change the world needs by helping others," is my motivational motto. So, what makes me an extraordinary engineer? Maybe nothing … I'm just a simple engineer sharing her story, but if one person gets inspired, then my job here is done!!!

Growing Up and My Educational Journey

Let's start with a little bit about me. I'm an only child. As I mentioned before, I grew up with my grandparents and my mom in Mexicali, Baja California, Mexico. Basically, I'm a "border girl" from Mexico. As an only child, the first granddaughter and the first niece, I got a lot of attention from my family. However, at the same time, I also got the "role model" title attached to me. That is how my cousins saw me; a "role model." That is an amazing experience, but it is also stressful. I started working hard at a young age to set a good example for them.

From the time I was young, I was very creative. I put a lot of passion into all of my homework and assignments. I knew that taking great notes was my responsibility, and I took that very seriously throughout the pursuit of my education. At the same time, I challenged myself to see what I was capable of. Combining my creativity with my discipline was a key factor for me. I have always been the student that worked hard, did all of the homework, put in the extra effort, and made sure I got all the points correctly. Basically I'm a perfectionist. That helped me to establish a good reputation at school with my professors. However, I learned that trying to be "perfect" is not sustainable. Understanding that you can fail sometimes can be hard, but it is part of the journey. It took me a while to realize this.

Elementary and middle school were easy for me and I was always at the top of my class. I remember in third grade while I was playing "being at school" at home, I discovered how to find X in an equation. A few years later I realized that I had solved a real math equation.

High school was an amazing time for me to make friends and practice a lot of hobbies like jazz, ballet, and cheerleading. The last year of high school was hard for me because of my career path selection. In the final year of high school we chose a concentration, and I chose math/ physics. The most difficult class for me was computer programming. I remember learning JAVA and C++. Although it was the hardest class for me, I was always committed to becoming an engineer; so I got through it by sheer self-motivation. I had always dreamt about becoming an engineer. I was inspired by one of my uncles who is a civil engineer. I even have pictures of me as a baby "helping him with his homework."

As the only child in my family, my parents worked hard to give me the best education. I attended private schools in Mexico and took advantage of all they had to offer. I also received scholarships during most of my time in university. A critical moment for every university student's life is selecting your major, but I already knew that I wanted to be an engineer from early on. In Mexico, being an engineer offers good financial stability, so choosing to be an engineer was an easy choice for me.

The Ballet Dancer

Other than engineering, my passion has always been ballet dancing. I started ballet when I was 5 years old, and have practiced ballet for more than 20 years. Ballet has been a safe space for me since the time that I was little. When I dance I forget about any stresses in life. I have always been a realistic person, even when I was young. Therefore, I knew that ballet would just be a hobby for me. I never really saw myself as a professional ballet dancer or a ballet teacher, but I continued practicing ballet for many years because it gave me a lot of peace and helped me with two very important things:

1. Discipline
2. Being a Team Player

Ballet is one of the harder exercises you can do. It pushes your body to extreme levels, and also requires that you train the mind, because everything must be done with grace. You must be conscious of every step and do it within the right form and with the right strength. To have this combination of body and mind is what I simply love.

Near the end of every year, we had dance recitals and I remember that during the presentations, there was a certain set-up time to change dresses between dances. Doing this quickly and within the specific time frame was just pure art. I have so many fond memories of those times.

Ballet gave me the discipline that I needed to have structure in my life, be organized, and be focused, which helped me get results.

Consistency is a key element in anything, and ballet helped me create that in my own way.

Ballet also helped me become a team player. There were so many team choreographies and everything needed to be orchestrated in synchrony with the team. You need to be able to trust all the dancers and be cohesive with them.

In my happiest moments and in my darkest moments, ballet always brought out the best in me. I will always be grateful to my family for supporting me in my hobby; every dress, every rehearsal and every afternoon after class. It was not easy, but it's amazing to go back and see all of the great things that happened in my life because of ballet.

University and the Industrial Engineering Journey

Let me tell you more about my journey to choosing industrial engineering. When I was in high school, especially during the final year, the school started to push students to think of life beyond high school. Students could have concentration classes as well as special classes in particular areas. I began to be more realistic about life in general, and even though I was not the best or the most natural fit for STEM (Science, Technology, Engineering, and Math) type classes, I chose to have my concentration be in Mathematics and Physics. This was because of my desire to study engineering, and that was the best path to achieve this.

Once you choose a major things get interesting, because you don't know what engineering disciplines are available and what the differences are between them. My first choice was civil engineering, since this is what my uncle had studied. I kept my eyes open for all of my options.

I was fortunate to be at a school which was a university and high school, where there were a lot of workshops and activities available to high school students, which taught them about the majors being offered at the university level. In one of those workshops, I learned about industrial engineering. This felt like a big "Eureka!" moment. Instantly I knew what I wanted to study. I am so glad to have been able to study at CETYS University.

I remembered reading that an industrial engineer is a "scientific administrator" and, for some reason, that caught my attention. After reading more about it, I knew that would be my dream career! Focusing on creating processes, working with people, statistics, being able to organize layouts, introducing visual management, and finding root causes of problems, were all things that I was interested in doing. Of course, at the age of 17-years-old you don't fully understand what you are getting into, but I can say that I made the right choice! Growing up at the border also helped me also to finalize my decision. Mexicali is a cluster for a lot of industries, and as an engineer it would be easy to find a job.

University was amazing. I was like a sponge, ready to learn and ready to contribute. I got a 75% scholarship for most of my time in university, but in my final year I got a 90% scholarship by Gulfstream. This scholarship was a blessing for my family and me. I applied for the scholarship which was only for top students and did several interviews. The application committee even reviewed the financial situation of each family in order to select the best candidates for this opportunity. At that time my family was struggling, so this scholarship was a relief in so many ways for us.

Industrial engineering in Mexico is a 4-year major. The first two years of study are common ground, which teach the foundations of engineering, and offers classes to learn about your field. The 4th semester is when the good part starts because it is the time to focus on your major (industrial engineering). I remember being happy about having more classes and specifically, about the subject of industrial engineering. One of the first subjects that I learned was Lean Manufacturing. That was a key moment for me because I realized that was something I wanted to study more in depth. Value stream analysis, 6S, and problem solving were all fascinating to me. As the semesters continued, I learned about many more subjects like Design of Experiments, Simulation, Quality, and many other great subjects.

I have never doubted the choice I made for my major, but I know that there are people that study industrial engineering because it is the "easiest" type of engineering. I want to explain that being an industrial engineer can also be challenging. You need to be versatile, understand the human factor in your simulations, understand that you are not the technical person, and at the same time, you need to be able to contribute while analyzing a process. You need to know statistics and talk with data, understand the big picture, and also see the small opportunities that can make incremental improvements in your process.

One should not study industrial engineering because it is the "easiest" one, but rather study industrial engineering to be the change and help others to have better processes and strategies. Don't ever choose something because it is the easiest as this will not benefit you in the long run. People who choose the more simple route often end up dropping out of university. They tend to not enjoy it or lose interest in it. If you choose it on your own terms, any major is the right choice because it will keep you busy and entertained.

During my time in university I encouraged myself to participate in different engineering events that helped to build my skillsets, in addition to the technical ones. I was part of the engineering society of my school, and during that time we organized benchmarks for different companies. One of my most memorable events was visiting the Google Headquarters. Additionally, I made several trips to local industries in Mexicali. I was also part of "Proyecto Ingenieria," which was a day to showcase projects from all of the engineering students on campus.

My proudest moment was when I participated in Trascendencias as part of the marketing team in 2011, and then as the General Coordinator in 2012. Trascendencias is an engineering conference that is organized by students for students. It is a summit where we organized workshops, lectures, and events for students to learn. I'm proud to say that I was the first woman in 28 years to coordinate the event. We gathered more than 300 students from the different campuses of CETYS Universidad (Ensenada, Tijuana, and Mexicali in Baja California). This event helped me to understand the big picture and the challenges that people face. I had an amazing team who supported me, and our main topic was "Challenging the Status Quo."

Selecting a major is not easy and I recommend doing your own research. Currently, there are many tools available to learn about different opportunities in universities. It is much easier today to find out about STEM programs in all places and I encourage you to use these new tools, and also do your own research. Additionally, being able to participate in all the programs that were available helped me to embrace the engineering life. Trust your instincts!

The Door to the World

One thing that I never imagined myself doing was to be an international student. During my time in university, I was lucky enough to have been involved in three different student residencies, in three different countries; India, Canada, and Austria.

Let's Talk About India

The first overseas trip I took was to India. Yes, to India! I spent one month there with 15 other students from my university. It was one of the best experiences of my life! This trip opened my eyes tremendously, and I will always be grateful to my school for offering this type of program. I am also grateful to my family for all of their effort and support in sending me abroad. One funny thing was that, even though I grew up near the border, this trip required me to get a visa.

We traveled to the Maharashtra District, where I visited Pune, Thane, Mumbai, Goa, and Lavasa. The program in India was focused on lectures about finance, marketing, economics, management, and engineering. I was able to visit TATA Motors, a steel company, and the Consulate of India. This is the trip that stole my heart. As a student you are never sure which type of work you will end up doing, but I have been lucky enough to work in international companies and with people from India. When I share that I know their country, this instantly creates a connection, and that is one of the most amazing things that I gained from this trip. Even after so many years, I am still able to remember this trip and create new experiences because of it.

Let's Talk About Canada

My second trip was to Vancouver, British Columbia, Canada. I know my strengths and my weakness well, and one of my weaknesses is languages. Speaking and reading English has always been hard for me.

It's crazy because I've had classes in English beginning in elementary school, and I have embraced it and tried to do my best. It does not come naturally for me. So yes, writing this chapter has been a real challenge for me.

Because of my struggle with English, I spent a summer abroad in Vancouver to learn business English. It was an amazing experience in an amazing city, and with amazing people. I was able to get to know people from many different countries during my stay in Vancouver. I lived in what is called a "stay home," where a family took me in and cared for me of me during my stay. It was an amazing new experience for me, and I have treasured all of these moments in my heart. One fact that is not much fun is that on this trip I got hit by a car during my second week there. This was not a nice experience, but after the accident I decided to continue on with my English program. All of my friends and my "stay family" supported me so well during this trip. Life is full of many experiences and you must learn to embrace them all, the good and the bad.

Let's Talk About Austria

My third and last trip was to Linz, Austria. I spent a summer there to participate in a marketing class with more than 15 students from my university. This was similar to the program in India. While you are in university you must take extra classes that are not necessarily related to engineering or your major. I took a chance with this trip and chose a marketing class, and I couldn't have been happier with the results. During this trip I was able to visit the BMW Motors facility, and to this day I'm still impressed by all the processes that I saw. That was a confirmation that I chose the right major for me!

The Internships

At my university they pushed us to take internships after the 4th semester and changed the class schedule to the afternoon to allow students to have the morning "free." It's not really free time because students are busy doing homework or studying, but it gave me time to start doing internships. Where I work, the companies understand this about students, and they usually have their internships during the morning hours.

I was really excited about doing internships. My first one came in 2013 at Bosch, in the business unit dedicated to power tools. I was an intern there for 11 months and it was a great experience for me. It was the first time that the concepts that I was learning in school became a reality. Also, something key from this internship was that I would have my first boss ever. I remember that he would explain all the concepts and processes to me in a very detailed way. I was responsible for collecting the continuous improvement ideas from the operators, and attended the daily production meetings.

After my internship with Bosch, while I was still a student, I decided to do a second internship. Internships usually last 6 months and students can take as many as their time permits. At that time, a friend of mine was finalizing her internship with an automotive company, and she asked if I was interested in the position. Of course, I said YES! I prepared my resume and she vouched for me. I got the interview, and I can still remember the questions that the manager asked me. Thankfully, I got the internship job.

The job was with Honeywell Turbo Technologies, and I had the best boss ever! She became my mentor, and during this internship I discovered my passion for "continuous improvement." I will be forever grateful for this opportunity. Additionally, after finishing my internship, my manager worked things out to be able to hire me as a Manufacturing Specialist. This is where my journey in engineering began. During my time as an intern, my boss gave me a lot of exposure. I conducted a lot of training activities that reached more than 800 employees. These trainings helped me to grow a lot.

I absolutely recommend gaining exposure to the industry while you are still a student. This will help you build your path and understand the big picture, while also connecting the dots for all there is to learn at school. This will also give you some advantages and experiences to be ready for the "real world" after graduation.

The Work Journey

Once I achieved my goal of graduating with an engineering degree, and I immediately moved onto my next goal ... to start working! I got my first job offer at Honeywell, in the Automotive Business Unit after my internship. I was fascinated by this industry, the processes, the machines, the procedures, everything! Helping others improve their processes is the reason I joined continuous improvement. Thankfully, because I had a great boss (I will be forever grateful to her), I learned a lot and felt very confident in my first job, although I had just graduated a few months earlier.

After some time, I moved to another Honeywell facility in the Medical Devices Business Unit. I joined as Project Management Specialist supporting operations, and one year later I was given the opportunity to become a Continuous Improvement Site Leader, at just 24 years old! It was a crazy time for me and I felt so lucky but, of course, I needed more preparation. I invested many extra hours to ensure I would be good enough for this new role. Being young and a woman can be hard in the industry. I worked hard to demonstrate that I deserved my position. Also, to help with my new role, I decided to work towards a Master of Science in Engineering, with a focus on industrial management. I wanted to ensure my knowledge and support would be better for the type of work as well as for my co-workers. During my time with Honeywell I started to travel to other factories in Mexico and the U.S., due to my work in assisting with maturity assessments, workshops, and trainings.

Since I like to focus on continuous learning, I started the Lean & Six Sigma Certification journey. The first certification I received was getting my green belt so I could learn more about it, and also apply my knowledge into a cost-saving project. After that, the nature of my role was to go to the next level of certification, which is Lean Expert. I participated in four weeks of training in different cities for a specific related project. Next, I went for the black belt which was the same process; 4 weeks of training and a project that took me a year and half to complete due to the complexity of the implementation.

The last training I did was the Lean Master Training. This was another 4 weeks of training, and it took me around the globe. To be part of this training I had to participate in tests and interviews, and I needed the sponsorship of my Business Unit Leader. As part of this training, I was able to travel to the U.S., Mexico, China, and Romania. I share this experience to showcase that when you are a committed person, you will create your own opportunities to succeed in something that you are passionate about. I will never forget all the great exposure, knowledge sharing, and successful activities that I participated in, but most importantly, the amazing people that I was able to meet throughout my career.

After some great years in my first real job, I decided to close that chapter in my life and pursue a new opportunity at another great company, Collins Aerospace. I enjoyed my time with Collins and was also able to travel. I supported the Continuous Improvement Program and had great co-workers. I was really fascinated by the importance and focus on quality to deliver products that we can trust while we are in the air.

Something I tried to focus on during my time with Collins was my "work-life balance." During that time, I did something that was out of my comfort zone, and became a part-time professor at my alma mater for a year. The best way to enhance your knowledge is to share your knowledge with others. This phrase is totally true, and it was very fulfilling to be able to share the things that I know. Being able to see the students apply their new knowledge in their projects often left me speechless.

Sadly, my time with Collins was short because I got married and moved to the U.S. I got married "JIT," *Just in time* before COVID hit. We had a very beautiful wedding in Mexico with hundreds of guests *(Mexican weddings are big weddings haha!). This* is how my work journey in Mexico ended. I had mixed feelings about all the changes that were happening in my life, but life is full of surprises.

From Mexico to the U.S.

I never imagined living outside of Mexico, but life is funny. I ended up moving to San Diego, California in April 2020. My husband is also Mexican, but he had a work opportunity in the U.S., so we decided that once we got married I would be the one moving. Growing up near the border helped me with cultural barriers, but my biggest concern was all the hard work I had done in Mexico. When I arrived in the U.S. I thought that I would have to start my career over from scratch, and that made me sad for a few months while I was waiting for my permanent residency. I was surprised that I was able to continue building my career with the same or better opportunities than I had in Mexico.

The day I got my work permit, I was ready to apply for jobs. My resume was ready, and I watched many videos about how to conduct interviews. I was set to achieve my goal. I worked as a consultant for a few months while waiting for "my dream job." I would like to say that I was lucky to find a job so quickly, but I think my luck is something I created by investigating and not being afraid to ask questions. Never limit yourself. I remember that one of my biggest concerns was not my technical knowledge, but rather my accent. I was so nervous, I recorded myself on my cell phone to practice my English.

My first job in the U.S. was at General Dynamics. It was funny because I think being bilingual helped me. I had a lot of coworkers that were Latinos. Another interesting thing about General Dynamics is that Gulfstream is part of that Company, and Gulfstream was the one that gave me a scholarship of 90% during my last year of school. I felt like working at General Dynamics was a moment for me to give back. It was destiny. It was meant to be and I loved working at General Dynamics. It had a very different focus. I learned a lot and I have many great memories of my time there.

Currently, I'm an Operational Excellence Manager at a biotech company called Illumina. Illumina is a leading developer, manufacturer, and marketer of life science tools and integrated systems for large-scale analysis of genetic variation and function. This was a 180-degree turn in my work. I changed from the aerospace defense industry to the biotech industry. Every industry has something in common whether you are building a car, a plane, or a sequencer; everything is going to be made by following a process. As an industrial engineer, I focused on understanding the steps to those processes. This is why I have been able to work in many types of industries, and it is why I love my career.

My inner child could never have imagined achieving my dreams, one by one, little by little, but with a lot of hard work. Changing countries can be difficult, and sometimes I feel like I'm in limbo between Mexico and the U.S., but again, I am always grateful for all of the opportunities I have had. I'm really happy and ready for more challenges.

Trust Me, I'm an Engineer

I'm happy to say that this year is my 10-year anniversary of work experience. I waited many years for this moment. During those 10 years I focused a lot on my career. Sometimes I made my job my first priority in my life in order to achieve my goals, but now my advice is to try to have a good work-life balance. This is something that I am still working on.

I have been able to train more than 2,500 employees, participate and lead more than 200 Kaizen events, and mentor 45 six sigma greenbelt and black belt candidates. I have supported and led our annual goals deployment, helped organizations apply different lean and six sigma concepts, and even helped with internal and external audits. I have spent endless hours doing Gemba Walks and process mappings. Most amazingly, I have been able to visit about 60 facilities around the world; from Mexico, to the U.S., China, Romania, India, Austria, and Germany. How crazy is that?! All I can say is that it's been an amazing journey! When you find something that you are passionate about, you will enjoy almost every second.

I don't want to romanticize work. It can also be stressful. I have cried many times. I have failed many times, and maybe for some people I may not be good enough, but the key is being yourself, care about what you do, and how you impact others. Hard work doesn't guarantee success, but it improves the chances.

The Next Steps

Engineering has been the greatest decision in my life. I enjoy being able to help others, have stability, be a team player, help on alignments for simple and complex problems, and help my family. It is a dream come true.

Since I moved to the U.S., I have done my best to be a mentor and participate in different activities that motivate other girls to pursue STEM careers. I was part of the Latina Engineer Conference in 2022 and 2023, and I was able to mentor four Latinas in 2023. Of course, I tried my best to share my knowledge with my teams during these years. I have helped more than twelve interns and four engineers.

Being an engineer has been a game changer for me. It would make me very happy if my story inspires at least one person to pursue their dreams. I am a hard-working girl and I will always try to do my best, and go the extra mile. I believe that going above and beyond is what makes the difference.

Lastly, I would like to give a few more tips: always be a person that can be trusted, ensure that you collaborate deeply, and show that you truly care.

I am grateful to all of my family (especially my Nana, Tata, and Mom), friends, and coworkers that became family, and of course, my husband who has been my #1 fan in everything that I do. It doesn't matter how crazy our schedules become, "*Families are the compass that guides us. They are the inspiration to reach great heights.*"

Lastly, I just want to say thank you to Justina for creating this amazing book and putting together real stories from real engineers. It is… extraordinary.

With love and openness, Grethel.

Connect with me:

Instagram: TheKaizenWave
LinkedIn: Grethel Cabuto

LinkedIn

Instagram

The Accidental Engineer who Persevered

Coauthored by Lori Kahn

I never saw myself becoming an engineer, but here I am a Systems Engineering Manager at Lockheed Martin Space, where I've worked since 1996. My path to Systems Engineering is a unique journey that found its way there from a deep passion for the space industry.

Growing up in Houston, Texas, I was an only child of divorced parents: a business-savvy stay-at-home Mom and a payroll accountant Dad, neither of whom completed college. I didn't have any role models in STEM (Science, Technology, Engineering, and Math) fields. I played with Barbies, loved games that challenged my mind, and excelled in my math and science classes. In the early 80s, the Space Shuttle was captivating all with the possibilities of discovery, and the potential to see a permanent space station orbit the Earth. And with the Voyager flyby passing Jupiter and Saturn, I became curious about Astronomy too.

In junior high, when my friend Katherine asked me and another friend to go to Space Camp in Huntsville, Alabama, I was intrigued and asked my mom, who agreed to send me there. The three of us had a blast! I returned to Space Camp (for their opening week of Space Academy!) after my junior year in high school, where I developed a desire to become an astronaut and wanted to do experiments in space (the Payload Specialist track). That curiosity grew further while volunteering at the Houston Museum of Natural Science with the Challenger Center, a mini–Space Camp experience for school groups established in honor of Christa McAuliffe, the first teacher in space who, unfortunately, lost her life in the Challenger disaster in 1986. At the Challenger Center, I gained two female role models in STEM; one who ran the Planetarium and its outreach and another who ran the Challenger Center that I worked with closely. My volunteering at the Challenger Center eventually turned into a paid job over the summers, before and after my first year of college.

In exploring the Payload Specialist track at Space Camp Level 2, one of the camp counselors helped me set up an alignment of optical lenses where we pointed a laser through it to see how light diffracts. That experiment hooked me on science, and I eagerly anticipated taking Physics my senior year. Unfortunately, I was discouraged when it didn't come easy for me like my math classes. I really struggled to understand the concepts and had to get a tutor for the first time to help me get through the class. I still managed to get a B, but my confidence was shaken, and I doubted my ability to go further with it.

Choosing a college was simple and easy for me compared to what it's like today. I applied to one school early decision, which I got into, and that was it. I chose Smith College, an all-women's liberal arts college in Western Massachusetts. Smith appealed to me the most after visiting several college campuses as a junior. I loved Smith's beautiful campus, that it was part of a 5-college consortium, and that there were no core curriculum classes (such that I could avoid taking my least favorite subject, English). I was leaning towards majoring in Astronomy, but I wasn't ready to take Physics again, and wanted to explore other subjects during my first year of college.

Once I took Intro to Astronomy with Dr. Richard White, who would eventually become my advisor, I was highly motivated again and decided I would pursue Astronomy as my major…which meant taking Introduction to Physics as a sophomore. It wasn't easy, especially when it meant I would have to take 4 classes at 4 different colleges in order to pursue the academic track I wanted. In order to take 2nd year Russian at Smith, I had to take Intro to Physics at Mount Holyoke, 2nd year Astronomy at Amherst (since it was a 5-college class), and that left Linear Algebra which I took at UMass. The bus schedule supported my class schedule, but I was strongly advised against it by my original advisor. I was determined to make it work and I did. I studied Russian all 4 years and did a junior year semester abroad program with Dartmouth College, that allowed me to take Physics at Dartmouth in their winter quarter and go to Moscow in the spring quarter of 1991. Going abroad, even for a short time, was an experience that helped me learn about another culture and see things from their perspective.

Another defining moment for me in college was joining Students for the Exploration and Development of Space (SEDS), which included an international network of students who desired to learn more about the future of space exploitation. I went to conferences at MIT, Toronto, England, and Florida where we had hoped to see a shuttle launch, and became the president of the chapter at Smith in my senior year. I had an astronomy research internship at Carnegie Observatories, part of Caltech's Summer Undergraduate Research Fellowship (SURF) program, that I found out about through my SEDS network.

Spending the summer of 1991 in Pasadena, California, allowed me to work with an outstanding female astronomer, Dr. Wendy Freedman, and go with my SEDS friends down to Baja Mexico to see a total solar eclipse. I graduated from Smith in 1992 with a double major in Physics and Astronomy (there were only three of us), an appreciation for cultures other than my own, and a love for the space industry. The next step was to get my PhD in Astronomy...or so I thought.

I applied to multiple graduate schools for Astronomy, however, I was only accepted to Arizona State University (ASU). I decided to defer that for a year in order to attend the World Space Congress in August 1992 with fellow SEDS members. I spent my summer after graduating from Smith living in London, doing astronomy research at Queen Mary and Westfield College, as well as traveling within the country. I spent the next year retaking some Physics classes at UCLA while continuing my astronomy research with Dr. Freedman on variable stars at Carnegie Observatories.

As a graduate student at ASU, I became more proficient in Physics, but after 2 years of coursework and an entire summer studying for the comprehensive exam, I didn't pass. I was faced with a tough decision. Do I keep studying and take the exam again (in one year), or write a thesis and graduate with a Master's degree? I chose the latter option - to write a thesis using observation data on standard stars that my advisor provided me. I also came to the realization that my future as an astronomer would depend on grants which are hard to come by, and rarely sufficient without also teaching at a college or university.

While I loved doing astronomy research, I didn't love teaching a class full of students, based on my experience as a Teaching Assistant during my second year at ASU. I wanted to follow my passion for the space industry. I checked results from my experience in academia, and decided that I needed to make a change in order to pursue a new career path that was better aligned with my passion and skill set. One of my Astronomy professors at ASU connected me with someone at Lockheed Martin Missiles & Space, who put me in touch with the Systems Engineering department that was hiring in 1996.

While I'm sure I was exposed to engineering through the many SEDS conferences I attended, it wasn't until I got offered a job in Systems Engineering that I became curious about it as a career. After getting the job offer, I promptly visited the engineering department at ASU to find out more about what engineering was. I moved to Sunnyvale, California, in 1996 for my first full-time job.

I joined the International Council on Systems Engineering (INCOSE) which had a local chapter and monthly meetings, so I could meet other Systems Engineering professionals and learn more about how they applied Systems Engineering at local companies. I learned as much as I could about Systems Engineering from INCOSE materials, and from my coworkers on the job. My early assignments allowed me exposure to early-phase NASA and Military space programs, where I was involved with developing spacecraft specifications and space vehicle concept of operations, before taking a 1-year stint on the Solar X-Ray Imager program from the point of build, test, and integration of this solar telescope onto a weather satellite with other instruments. These first few roles helped me gain more knowledge about Lockheed Martin as a whole, make meaningful connections about Systems in these different roles, and explore different perspectives from working with different teams. I also met my husband, Dan, through a Lockheed Martin social group. We got married in 2000.

In 2002, I transitioned to a Systems Engineering role within Special Programs-Remote Sensing (SP-RS). In order to better learn about spacecraft engineering I applied to Stanford's certificate program for Spacecraft Design and Operation Proficiency, which was taught by Professor Bob Twiggs. This program was a cohort with other Lockheed Martin engineers where we worked as a team to build, design, and test a CubeSat in one year. I learned the basics of spacecraft engineering from Space Mission Analysis and Design by Wertz and Larson, a common handbook on everyone's shelf in the office, and hands-on learning in the lab where we soldered circuits together. During 2003, I was also pregnant with my first son which presented its own set of challenges, but like all other challenges, my determination to get as much out of the experience worked out. Coincidentally, three of my male teammates were having babies too, so we affectionately named our CubeSat RABBIT (Rapid Application & Basic Bus Integrated Testbed).

After 10 years at Lockheed Martin, I began to understand better how a defense/aerospace company works, and what types of things systems engineers do at Lockheed Martin. Thus, I became more confident in my skills. I honed my requirements systems engineering expertise in SP-RS, and set my sights on the leadership track. While I was a group lead, I received some feedback about my interpersonal skills that made me doubt my ability to lead people. This set me back on pursuing systems leadership roles, but I still took action to improve in this area.

I had recently explored working in risk management with suppliers, which I enjoyed, so when asked to take a role as a product systems engineer on a subcontract management team, I said yes. This role let me work collaboratively with a supplier on updating their specification, and eventually the verification of those requirements.

I had my second child in 2007, and both my LM and supplier teams were extremely supportive. I appreciated the subcontract manager role so much that I decided to pivot from Systems Engineering into Subcontract Management. In order to do this, I needed to understand more about subcontract management, so I took a series of training courses and enrolled in a mentoring program, where I received a seasoned Subcontract Manager for a mentor. Because of the mentorship I received and my genuine interest, I was offered an opportunity to take on a challenging subcontract management role for an advanced star tracker, which consumed me for the next 3 years.

I learned a LOT about managing a technical team through many complex issues, and developed a solid relationship with my Program Manager counterpart. After delivery of the first star tracker, I was hoping for another subcontract opportunity, but there was not one available and I almost lost my job. It was quite a low point for me, but I bounced back as I always do. I quickly found another role in Systems Engineering as a Systems Integrator (similar to the Product Systems Engineer role), and I kept applying to subcontract manager roles outside of my line of business as they became available. I learned how to improve my interview skills with each interview, but unfortunately, I kept getting rejected.

In parallel, I spent time developing my leadership skills by taking an active role within our Women's business resource group, Women's Impact Network (WIN), by starting a Lean In Circle inspired by Sheryl Sandberg's book, Lean In. I became the first Space WIN chair in 2016, where I managed the activities and budget for WIN site leads from 10+ sites at Lockheed Martin Space. I later held a role on Lockheed Martin's corporate WIN leadership, where I coordinated with WIN leads from five business areas in an effort to share best practices and enhanced collaboration. I also joined the Society of Women Engineers (SWE) in 2014, and became active by making presentations at their local and annual conferences on diversity and inclusion, and STEM outreach topics.

From the connections I made in WIN, I met a subcontracts director who offered me a stretch assignment as subcontracts lead on a proposal. Winning this would put me in a good position to be the preferred candidate for the subcontracts lead manager role. While this opportunity helped expand my connections within the subcontracts department, and I learned more about subcontracting at a higher level, after two iterations of proposal development, we never got to submit it. When I was presented with an opportunity to get a promotion as a lead systems engineer for a new proposal in Military Space, I stopped looking for subcontract management positions.

Returning to Military Space after a long gap, I discovered that their knowledge and expertise in Systems Engineering wasn't at the level I was used to within SP-RS. I found myself as the Systems Engineering expert on a proposal, and while we waited for the proposal results, I worked on another program where I got to help develop their System processes. Unfortunately, we didn't get the new business and that meant I wasn't getting the Systems leadership role I desired, but the experience boosted my confidence in Systems Engineering expertise.

After years of successful leadership experience within WIN, and confidence in my technical work, I decided to apply to Systems Engineering leadership roles within SP-RS. I didn't get the first one I interviewed for, but I was encouraged to apply for another opening and got the second one. I've been in a Systems Engineering and Integration Manager role since 2020, where I led 10 Systems Integrators through Preliminary and Critical Design Reviews with their respective spacecraft product teams. My role is still a bit technical since I'm guiding both early and mid-career Systems Engineers to use the right processes and tools to be successful in their roles. At the same time, I've learned a lot about managing people, and found that I really like that aspect, especially where I can help influence others' careers.

I've continued to get more involved in SWE and STEM Outreach. Since 2018, I've been a mentor for the FIRST Robotics Team 2813 Gear Heads at Prospect High School, where my sons attended and have both been active with the team. There was 1 girl and 1 female mentor on the team when I first volunteered. I helped the team increase their outreach and affiliate as a SWENext Club in order to encourage more girls to join. The team is now made up of 35% girls and a few female mentors. I've also been involved with leading Invent It Build It, which reaches 150+ middle school girls at SWE's annual conference. SWE has taught me the importance of being a Role Model, and sharing my story with other girls to encourage them to go into STEM careers.

Another organization I've been supporting since 2014 is TechWomen, a U.S. State Department-funded program which brings 100+ women from Central Asia, Africa, and the Middle East to Silicon Valley (and now Chicago) to collaborate with professional mentors, cultural mentors, and impact coaches over a few weeks. I've supported as a cultural mentor for many years and more recently as an Impact Coach for Tajikistan, Uzbekistan, and Morocco.

In 2019, I participated in the delegation trip to Uzbekistan where I shared about my career and how one can get involved with robotics as a teaching tool for inspiring STEM in young students. This program has connected me with other like-minded women in technology, who are just as passionate about STEM outreach as I am. I've also learned a lot about their cultures as they have about mine.

In 2023, I was honored to receive both the SWE WE Local Engaged Advocate and SWE Advocating Women in Engineering awards. In addition, to honor my mom, who was the biggest supporter of my career, and my love for SWE's mission, I set up a SWE scholarship in her name, the Melanie Clampitt Memorial scholarship. SWE handed out the first scholarship from this fund in 2023, and I look forward to helping many more young women receive scholarships in order to pursue their career in engineering.

Connect with Lori:

…I am a ~~Jedi~~ Mechanical Engineer, Like My Father Before Me

Coauthored by Alexa Zsofia Vas

Growing up in a "petrolhead" engineering family has affected my future in the best possible way. I spent most of my childhood in the garage repairing family cars, household items, or just tinkering in general.

About my family, starting with my grandfather: he was a technical genius who achieved a lot in his life. Originally, he was a technician specializing in chemical engineering. He later studied management. After his compulsory military service (where he was, not surprisingly, a technician and a driver) he worked as a driving instructor. Later, as he went back to his original profession, he founded a company that cleaned the hazardous material tanks at the Chemical Works of Gedeon Richter Plc. Professionally speaking, he did a LOT. He always taught me about material science and machinery. He is the main reason I chose machine design and production as my major at university.

There is one more reason I wanted to be an engineer so badly, and that reason is my father. He is also a mechanical engineer (fun fact: we both got our degrees at the same university, and in the same subject and specialization).

My father had a tough childhood. He had to take care of himself from the age of 14, so he couldn't afford to enroll in a university. He started working as a simple production line worker at Suzuki in Esztergom. It was at the age of 40 that he was able to start university, which he completed as a correspondence student, while working and having to deal with 3 children (aka my siblings and I). He always knew that lean production/management was his passion, so he got a second degree in this field.

I'm extremely proud of him, because during university I worked as an intern as well (we'll get back to that later) and it was really hard. I couldn't imagine how hard it must have been for my father. I vaguely remember as he was studying for his exams until dawn, he would sleep a bit and then head to work. But he made it, and I'm so proud of him, because all of it was a superhuman achievement. I always mentioned my dad in interviews because of his dedication to mechanical engineering and what he went through to fulfill his dream. Once someone shared a post or story on Instagram with the hashtag #mydadisbetterthanyours and I instantly knew it was directed at me. Yeah, girl, your father might be better, but mine is the only man who has never given up on his dreams and he did what he wanted most: became an engineer. He may not be a rich businessman, but he gave us his all. His perseverance also motivated me to achieve my own dreams.

These two men had a profound impact on me. From an early age I was interested in cars and everything that was labeled "not girly." I watched F1 & Rallye with grandpa and Star Wars with my parents. I loved math, physics, chemistry, machinery, biology (especially anatomy), and so on. Because of this, I was called a nerd. I was basically an outcast in every class. But I got used to it, and I also had my supportive and also nerdy friends group. In high school I was bullied because of my "boyish" interests.

I went on to found the student ambassador program of The Association of Hungarian Women in Science (NaTE for short in Hungarian), which was a group for girls who loved STEM (Science, Technology, Engineering, and Math). Here I felt normal and understood again. I made new friends, and found my besties, Bia and Kata. We had a Saturday get-together almost every month where we worked on fun projects, listened to lectures in various STEM fields, learned programming, built a Mars Rover, or simply had fun while we were making plans for Girl's Day. Girl's Day was our biggest event. It is a career day meant for girls who are interested in STEM. Most universities and companies like Bosch, Audi, and Mercedes joined this initiative and wannabe STEM girlies could find out more about the company's profile, the projects that employees are working on, the qualifications needed to fill a position, and so on. As you can see, there are plenty of career options in STEM, and it's easier to make a choice if you know what a mechanical engineer actually does.

Aside from Girl's Day, we also had an annual event: the NaTE summer camp. Back in 2016 or 2017 at a NaTE summer camp, I was introduced to one of my biggest passions in my life: 3D printing. Later that year I went to Hungary's biggest career event, Educatio Exhibit (Faire) which was similar to Girl's Day, but the companies and universities were in one place with their own stands so you didn't have to visit them. It was here that I randomly found Péter Pázmány Catholic University's stand where a 3D printer was on display under a sign that said: "Faculty of Information Technology and Bionics." I was a tad bit surprised, because Pázmány is widely known as an institute for cultivating law and pedagogy, but it turned out that they have a faculty for something that was interesting to me. I went to their stand to ask for research opportunities as a high school student. To my surprise they said "yes," and gave me a contact to my future supervisor, who helped me prepare for TUDOK. This was one of the biggest competitions for Hungarian high school students.

Being that I was interested in anatomy, bionics was an interesting and new manner for me in which I could combine anatomy with 3Dprinting, programming, and electronics. From that point on I spent every Friday at the university with my supervisor (I had to get research permission from the principal in order to do that). We started working on a fun new project: developing a low-cost, 3D-printed robotic arm made especially for children. We wanted to make this product as simple as possible, because at the very beginning our plan was that the robotic arm would be completely free, and everybody would be able to make it at home from scratch based on the instructions and models we provided. We also wanted to add a little bit of spice, so we experimented with EMG sensors to connect and match the movement of the hand with the electrical signals generated in the muscles. By transforming these signals into numbers, we can give commands to the control system, and thus, the wearer would be able to decide when they want to hold something or when they want to let it go. Although we didn't have much time until the competition, we achieved what we wanted. I managed to reach second place at the semi-finals and later won a special prize at the Carpathian Basin Grand Final.

In 2018, after the Grand Final, I won the "Person of the year" award by Qubit because of my project. I also gained media exposure, through outlets picking up on my project and story; I was interviewed a lot. Even Forbes wrote about me! Unfortunately, despite the results and the principal's permission, some of my teachers began to spread the word that I was not doing anything, and that I was just a dumb mediocre student who could not have achieved such success. This took a toll on my self-esteem, and I still suffer from imposter syndrome to this day thanks to them. I still don't understand why all of this happened, because it did; I developed a thorn in my side. I wanted to show the world that I am not an average student and that I am capable of anything. I even lied to everyone about my career choice because I was afraid that they would mess up my university application. These teachers were on THAT level of pettiness. So, this is how I enrolled in engineering school, though not as a normal student, but because I set the bar high.

As I was deciding on my career path, I knew what mechanical engineers did, but when I started university my horizon expanded. I started my studies in a special form that is called cooperative education, or dual education. This means that students are working as an intern the very first semester along with their studies. This way, when you finish university you will already have 3.5 years of experience compared to normal students, who normally have a maximum of 1–1.5 years of an internship. I finished my degree in this educational form while working at one of the biggest German luxury car manufacturer's tools factories. You know, the one with four rings.

In the first two semesters we were taught basic locksmith skills, or how to write CNC G-codes. We learned to operate a lathe or a milling machine, and fundamentally acquired all the knowledge that would help us become excellent engineers in the future. Engineers who know the manufacturing processes by heart, and therefore, will not make banal mistakes in the design or in the manufacturing phase. Fortunately, I already knew a lot about these things, courtesy of my dad and grandpa. They were into old-fashioned engineering: technical drawings on paper, milling, turning, basic metalworking, vernier calipers, and that sort of stuff. What I learned there completely changed the way I thought about mechanical engineering.

In my first summer internship session and in my third semester, I got a taste of press tool design and production. This still laid with the traditional way I knew, however, it was at this phase where I learned how to use the CATIAV5 modelling software (still my favorite, Dassault Systèmes did a great job there), I learned about surface modeling, and I created my first 3D milling program. Later I moved to the area where I felt I could really fulfill my potential. This was the body shop part of Werkzeugbau (to me, it sounds better in German, haha) within that the robotic simulation department. There's an amazing feeling you get when you start doing something completely new, and it just goes as well as if you'd been doing it your whole life. Even if you make mistakes, it doesn't feel like a complete disaster. This is how I felt when I was doing robotic simulations. Everything was magical, I had never enjoyed something that way before. I felt that I found my passion and I could do this until pension. I loved being around industrial robots, and I learned a lot about the various bonding technologies of car body parts. I even had an amazing mentor, and amazing colleagues with amazing ideas.

A year and a half before my graduation I started working on my thesis with my mentor, who was a specialist in roller hemming. Roller hemming is a sheet forming process during which, in the case of a door, the edge of the outer sheet of the door is folded onto the inner sheet, thereby making the complete door more solid and the edges processed. This was performed with roller tool heads mounted on robots. There is a separate tool head for each car type and for the various body parts. Before folding the edge, glue is applied to the surface of the plate in order to make the connection more secure, and this also provides extra strength to the plates processed with this method. The glue is cured in a later process, making the hemming an insoluble bond. This process was basically the most difficult and required the most expertise, which is why I really liked it and wanted to write my thesis in this field. I also wanted my thesis to be a bit more extravagant and invent something new.

I decided to design an adjustable roller hemming tool head, which would suffice as a single tool head for each robot. This would reduce the changeover time because the tool head would be fixed, and the rollers could be changed automatically by the robot. Once I had the concept in my head, I started designing the tool head in CATIAV5. There's a metal 3D printer at our development center, so the idea was to make the initial part using the 3D printer instead of flame cutting. Moreover, the printed piece can be milled in the machining department of the tool factory based on the 3D model. After I finished the design, I started to arrange the production of the 3D-printed tool body and its machining. At the same time, I also started on the robot simulation.

The simulation tests were carried out on the bonnet of my favorite car, because it's a relatively large piece of work with a lot of steps, so it was perfect for testing the roller positions. The simulations went smoothly, and I could prove that this invention would be handy. Everything was coming together, and it was a challenging experience for me because I had to do all the steps that would normally be carried out by separate people. For example, the design engineer doesn't do the work of the simulation engineer and vice versa. But I learned a lot from the process, and I think for someone who is young and wants to work in this profession, it's an invaluable opportunity to try out various fields in engineering. You only get to know something pervasively if you work in it.

For those of you who are curious about what happened to my 3D-printed robotic hand project: during my university years, I founded a startup dedicated to this topic. We went to competitions and we did quite well, we even attracted investors. However, I had to admit that regular work and startups don't sit well together and I chose my job. At the same time, I was offered a job at the world's biggest engine factory (where I still work) and I chose the safe option. I would like to come back to this project later, if I have enough professional competence and time. I don't want to leave it hanging, but now is not the time. However, I didn't give up 3D printing, I deepened my knowledge about the subject. Now I create content on TikTok about 3D printing because I want more people to know about this manufacturing process.

I was still working on my thesis when I got the opportunity of a lifetime. As I previously mention, I received a job offer from the company's engine factory in the production technology engineering department. It was an uplifting feeling that, although I hadn't finished my degree, I already had a job offer in my hand. Of course, I accepted. I always thought it was important to get an internship as soon as possible during my university years (dual education was perfect for this), because I knew that this was the only way to get a proper job. Students who don't put the effort into their internships are less likely to get a job offer. So dear reader, if this situation concerns you, I recommend you find an internship as soon as possible and spend as much time as you can in an industrial environment. You can fiddle around in university labs in your spare time, but that won't get you a job. Only persistence and a willingness to work will get you ahead.

At the end of 2021, I started working at the engine factory in the production technology department, which was my first ever full-time job. The engine factory produces a wide range of engines, each with its own area of expertise. For example, we made V6 Otto engines, R4 and R3 engines, as well as electric machines. My job here was to design the production technology for small cubic centimeter petrol engines (1-liter R3 and 1.5-liter R4 engines). Within the Production Technology Engineering department, I was moved to the Assembly department, where I became the engineer in charge of the Complete Engine department. The assembly department consists of two parts, the Complete Engine Assembly and the Basic Engine Assembly. In the assembly of the basic engine, the engine base is assembled: the crankshaft is installed in the block together with the cranks, pistons, cylinder head. The control and the valve cover are mounted on the engine. In the complete engine section, we mount the turbo, the engine cable, the dual mass flywheel, and the clutch. On the basic engine there are fewer people and more robots, and on the complete engine there are more people and fewer robots.

Since I came here as a rookie, I got the complete engine part, which I didn't mind because we had robots. My job was to design the production line machinery operations, organize the layout of the production line and solve the problems that arise during assembly. If necessary, we can also help shape the product during the development process, because sometimes a modification - which may not even be a major one, just a slight change in the design - can cause us unexpected problems on the production line that can cost us a lot. We can save the group a lot of money if we spot these and report them to the development department.

For me, it was quite a rocky transition into my occupation since I had been working in robotics and metalworking for years before that, and when I came to the engine factory, I only had some fundamental knowledge of internal combustion engines. So, I was pretty clueless in the beginning. Luckily, I had colleagues who I could turn to for help. They got me involved in engine assembly and explained how things worked around there. The job was quite stressful, and I often missed not being able to express my creativity properly.

In the spring of 2023, a once-in-a-lifetime opportunity came my way. I received a job offer from a German company. So after two years I decided to leave the four-ring engine factory and return to my roots. This offer purposed an immense challenge to me, even though I am professionally qualified for my future job. New country, new culture, new opportunities. I am beyond grateful that I was given the chance to move to another country, because I have wanted to leave Hungary for better perspectives for quite some time.

German is my first foreign language and Germany is the perfect destination for me on my professional path. I recommend going to another country for a while to everyone, either while pursuing a degree or already working in your career. In 2-3 years' time, you will gain invaluable experience, and when you go back home this chapter will be a huge head start for you.

I'm planning on a long-term stay in Germany, but I'm not ruling anything out. You must always be open-minded; looking for new opportunities that will take you further in life.

Connect with Alexa:

Questions and Answers with the Extraordinary Engineers

Taking Flight as an Engineer

Contributed by Bianca McCartt
GE Aerospace

Tell us about yourself?

"THOU SHALT FLY Without Wings," was a film about horses that I grew up watching as a kid at the Kentucky Horse Park International Museum of the Horse. Back then, I never expected *flight* would play such a central role in my life.

Until I was 16, I imagined my future career would involve the horse industry, either as a trainer or veterinarian. This was well within my comfort zone since I grew up on a horse farm in Kentucky. It was not one of the big, fancy farms owned by millionaires, though there were plenty of those around. My parents were hardworking immigrants to the U.S., chasing dreams of riding the world's best horses. My mother ran a small business training, teaching, and caring for the horses owned by others.

Throughout my childhood, horses were always part of my life, starting from when I was able to sit up on a horse's back, before I had learned to walk. By the time I was 5, I had a pony and by age 10, I had trained a few young horses myself. But as anyone who has been around horses can tell you, they are labor intensive and expensive to take care of.

I was excelling in my high school studies, which led to considering what other opportunities might be available to me. I was strong in most subjects (except athletics), and started taking as many AP courses as I could. As a junior, I was in the advanced placement Calculus class, and I found that I enjoyed solving problems. Our Calculus teacher invited a special guest from the University of Kentucky School of Engineering, Dr. Bruce Walcott, to present to our class. I had never considered engineering before that day. He shared with us some of the exciting work that engineers do, and assured us that if we were doing well in math and science, we could handle the engineering curriculum. He also mentioned that being an engineer meant job security, a good salary, and exciting career opportunities. I was sold! After doing what I felt was more than my fair share of backbreaking work in all manners of weather, I was looking for stability and work that wouldn't be too physically demanding.

As I started looking at engineering programs and talking to

engineers about what their work was like, my interest grew. A longtime client of my mother's was a mechanical engineer, and he took me on a tour at his employer, Lexmark. I remember seeing a 3D CAD assembly on someone's screen and I thought it was just beautiful.

After that I focused on mechanical engineering, and I was especially interested in automotive design. This was partly because I also enjoyed drawing, and felt that I could create awesome things as long as I could visualize them. I applied to the University of Kentucky (UK) Engineering program and was accepted. With the benefit of scholarships and financial aid, I enrolled and started my journey as a first-generation college student, in September of 1998.

I participated in the UK Society of Women Engineers' summer campus program for freshmen women engineers. As part of that experience, I met amazing women engineers, including the astronaut, Joan Higginbotham. At that point in time, she hadn't yet flown a mission, but it was incredible to me to learn that engineers could become astronauts. I was really inspired to discover the range of potential opportunities for engineers.

In the fall, when I started class, I was enrolled in the very first "Intro to Engineering" course that the university was piloting, to give all new engineering majors exposure to the basic principles of design and experimentation in their first year. We started off by working on team projects to invent, design, and build something real. The class was led by two instructors, one of whom was Dr. Sue Noaks. So, from the very beginning, I had great women role models. My academic advisor was also the university's Director of Women in Engineering, Sue Scheff. With her guidance and the support of UK SWE, I felt like I belonged in engineering very early on.

Why did you choose your field of engineering?

Mechanical engineering appealed to me because it has many broad applications, and you can see the mechanisms of the machines you work on. Mechanical engineers are part of the design of many of the products and systems we interact with every day, from cars, to aircraft, home appliances, heating and cooling equipment, energy and power generation, plus much more that enables our modern lives. I was most interested in creating things that move. After I made the decision, to pursue mechanical engineering, I learned that they work on an even wider array of systems and industries than I had initially realized, including many who are in different aspects of aerospace and manufacturing.

Where do you work and what are your responsibilities?

Shortly after I graduated with my mechanical engineering degree, I found a job at GE Aerospace (then known as GE Aircraft Engines) on the Edison Engineering Development Program. I have now been there for 20 years in a variety of different roles, always working with commercial jet engines at the Cincinnati, Ohio, headquarters campus. The common theme in my career has been that I am always looking for a problem that I can solve, and looking for ways to improve the products and systems within our company.

My current role is a leadership position focused on how to ensure our engineers are as effective and capable in their work as possible. This includes developing the process to hire new graduates for the Edison Program, as well as collaborating with external partners to build resources for engineering employers. This is a strategic role rather than a tactical technical one, and gives me the opportunity to positively impact the careers of many other engineers at GE Aerospace. This fits with our company purpose, "We Invent the Future of Flight, Lift People Up and Bring them home Safely."

Can you share some highlights of your career?

Throughout my career, I have had the opportunity to contribute to cutting edge technologies. When I was still a student, I founded the UK Solar Car team which designed and built UK's first solar car for the Formula Sun Challenge track race. This team has continued for over 20 years, building 6 generations of competitive solar cars, and giving the opportunity to countless students to practice their skills with a team.

At GE Aerospace, the first jet engine that I worked on was the

GEnx, which powers the Boeing 787. When I started in 2004, this was what was called a "paper engine" in that it existed only on paper at that point. I was part of the engine combustor design team working on a new technology to control engine emissions like NOx. It was really exciting the first time I saw my design turn into a real physical object which was part of an engine. It has been just as exciting every time after that too! The GEnx engine was the first new large commercial engine product for the company in over a decade, and I worked on it for about seven years, until it was certified to fly by the FAA.

I also earned a patent when I was part of a team that developed a new turbine blade inspection system, to replace a laborious manual process with a high-tech thermal imaging camera on a robotic arm. Several years later when I visited a manufacturing shop, I was delighted to find they were using the very first of these machines to go into production. Without revealing that I had any connection to this machine, I asked the operator running it how he liked it, and he started to brag about how awesome it was. It was a really proud moment to see that!

A more recent highlight of my career was working on the LM9000 gas power turbine engine based on the GE90 jet engine design. This new engine took advantage of the high efficiency of an aircraft engine to provide flexibility for power generation operations. One of the challenges of converting from an aircraft to ground operation is that the combustion must be much cleaner. To accomplish this, it requires a more complex combustion system that experiences significantly more acoustic noise than the flying version, so we needed to create a special acoustic damper device (or muffler) to mitigate those effects. This was the task I was given, and when I first started working on it, it was little more than a simple drawing of two boxes connected to represent a volume.

The difficulty with this concept was that you needed a wall between the two volumes that would be entirely enclosed, and somehow you had to connect it all together in a way that was inexpensive and could be properly inspected. It was a fun challenge to solve and be involved in learning all about acoustics, resonance, sheet metal forming, brazing, welding, inspection, and the advanced analyses required to validate the strength and durability of the design. I was able to come up with a creative and elegant solution to this problem and see it through to the engine test.

What was more challenging than you anticipated?

Of course, there were many challenges along the way through each of these experiences. Machines in general, and turbomachinery, specifically, are exquisitely complex systems. A big part of being successful in this field is being able to work well with a team, keep up with a ton of details, and always being willing to learn new skills. There were times when I doubted my own abilities or felt that others doubted me, especially when I made mistakes. But I have learned that everyone makes mistakes at times, and the key is to keep trying and learn from them. Sometimes I felt a lot of pressure to justify my engineering judgements to others with more experience and expertise than I had, but the reality is that no one knows everything that is necessary to build these products. It takes thousands of people contributing their knowledge for each machine that is designed, built, and flying around on an airplane. It took a while for me to realize that my unique way of looking at the problems and solutions was a valuable contribution to the team.

What do you wish you knew before you chose this career?

When I first started in engineering, I had an idea of design being about the way things look and the shape of the objects. I notice a lot of new engineering grads are focused on how to form parts, creating designs in CAD and 3D printing them to build something. While this is certainly a useful skill (especially in the field of industrial design), it only scratches the surface of what an engineer actually does. The important decisions in engineering are not only about the form, but also the fit, function, durability, cost, weight, safety, and how this object will fail at the end of its journey.

Safety comes above all, in that it must be safe for the person whose job it is to make it, safe while it is in use for its intended purpose, and operate safely throughout the total product life. This comes down to understanding the physics, materials, and processes that go into making the machines, and being able to validate that the decisions that are made while designing the objects can be validated to work in real situations.

Physically testing at each step of the process is prohibitively expensive when you are talking about a complex system, so you have to be able to use theory and make reasonable estimates through calculations. I also wish I had known how fascinating and fulfilling it would be to learn all of these things, and get a glimpse into the complexity that most people take for granted every day.

Can you share additional ways you promote women in engineering and/or STEM?

I have been active with SWE since college and into my professional career. SWE has brought a lot of value to me personally and professionally in the form of role models, mentors, scholarships, leadership skills, and insights into others' career journeys and industries. I am inspired by SWE's mission to "Empower women to achieve their full potential in careers as engineers and leaders; expand the image of the engineering and technology professions as a positive force in improving the quality of life and demonstrate the value of diversity and inclusion."

I feel it is important to give back to an organization that has given so much to me. I attend the SWE Annual Society Conference, volunteer as a mentor for other SWE members, and serve as a volunteer leader in the SWE Board of Directors. For me, this is part of a lifelong commitment to help advance opportunities in STEM (Science, Technology, Engineering, and Math) for women.

Connect with Bianca:

Sparkletronics

Contributed by Ayesha Iftiqhar
Senior Electrical Engineer at Lightship RV

Tell me about yourself?

HELLO, I'M AYESHA, a full-time Electrical Engineer dedicated to Clean Tech Solutions. In addition to my professional role, I'm a part-time Maker, STEM (Science, Technology, Engineering, and Math) Community Builder, and STEM Outreach Program Creator. I'm also an avid hiker and roller-skater. My roots trace back to Mysore, a quaint town in India, where I spent my formative years surrounded by captivating birds and animal sanctuaries. This upbringing instilled in me a profound appreciation for nature and its preservation.

Being the youngest of three girls during a time when having a girl child in India wasn't widely celebrated, I grew into a shy kid who sought solace in books. The Adventures of Nancy Drew particularly resonated with me, as I admired how a young girl could independently solve mysteries. My education took place in an all-girls school, an empowering environment where I witnessed the accomplishments of fellow female students in academics, athletics, and dance, which reinforced my belief that girls could excel in any field when given the opportunity.

My childhood fascination with science led me to devise experiments using everyday materials, such as creating a spectroscope with a cardboard tube, a fluorescent lamp, and a CD borrowed from my sister. The joy of hands-on work and crafting things became a constant source of happiness, shaping my decision to pursue this passion throughout my life.

Why did you choose your field of engineering?

I grew up lacking female role models in strong careers who resembled me. However, my perspective shifted when I discovered Dr. Kalpana Chawla, the first Indian woman to go to space. Her story ignited a passion in me to aspire for more in my own life. Dreaming of becoming an astronaut like her, I discovered that anyone could obtain an amateur radio license on Earth, and communicate with astronauts in space. The notion of conversing with astronauts captivated me as a child, sparking my interest in amateur radios and the associated electronics.

This newfound fascination became a driving force, propelling me to delve deeper into the world of electronics. Fueled by my aspiration to become an Electronic Product Designer, and driven by my inherent passion for crafting things from random materials, I embarked on the journey of obtaining a degree in electrical engineering. As a maker at heart, I aimed to merge my love for creating with my dedication to understanding the intricacies of electronic design.

To gain expertise, I worked tirelessly to move to the United States for a master's degree, recognizing it as the ideal place to pursue the work I envisioned as an electrical engineer. My studies focused on analog and digital circuits, embedded systems, and RF electronics. I aimed to comprehend every aspect of electronics integral to my goal. The journey, inspired by the remarkable Dr. Kalpana Chawla, has shaped my path towards a fulfilling career in the field of electrical engineering.

Where do you work? How did you get here?

I'm currently working as a Senior Electrical Engineer at Lightship RV. My journey here stemmed from a strong passion for addressing climate challenges, backed by a master's degree in electrical engineering. I started my career as a Junior Electrical Engineer at a company specializing in IoT-connected sensors for restaurants. Our focus was on creating systems to monitor and reduce water and power usage in industrial dishwashers and linen washers. This initial experience laid the groundwork for my career in electrical engineering.

Later, at Ryobi, I played a pivotal role in designing motor control units and battery management systems for battery-powered outdoor tools like lawn mowers and chainsaws. The tangible impact of reducing the carbon footprint of everyday household tools in the USA was incredibly gratifying. Growing more confident in my electrical engineering skills, I sought out challenges that stretched my creativity. This led me to a startup company that worked on fruit-harvesting robots for hydroponic growers utilizing UV disinfection instead of chemical pesticides.

Motivated by my love for the mountains and nature, I made the decision to join Lightship RV. Now, as a Senior Electrical Engineer,

I'm actively engaged in designing the first all-electric RV in the United States. This innovative vehicle not only facilitates nature exploration, but also contributes to keeping the air clean and nature undisturbed by harmful air and noise pollution, aligning perfectly with my commitment to sustainable engineering solutions.

What are your responsibilities?

My role as an Electrical Engineer encompasses various responsibilities throughout the design cycle. During the initial stages, I will collaborate with cross-functional teams, including product design, mechanical engineering, and software engineering, to create architecture-level designs. This involves tasks such as calculating power consumption, selecting appropriate sensors, motors, and controllers, planning wire harnessing, and minimizing emissions. Following the architectural phase, my focus shifts to translating these designs into custom circuit boards, where I design printed circuit boards. Once the design is complete, I conduct thorough testing at the board level to ensure functionality. Subsequently, I then integrate the boards into a larger system, collaborating with the software team to write firmware and perform comprehensive functionality checks.

The next phase involves extensive testing for compliance and reliability to ensure the verified design meets standards. Additionally, I work on developing production-level tests and test fixtures to ensure the scalability and mass production feasibility of the design. It's not uncommon for certain aspects of the system to undergo redesign, leading to iterations of these steps for continuous improvement.

Can you share some highlights of your career?

In my journey, I've reached various milestones, like establishing my small business. Being chosen as the Ada Lovelace fellow for the Open Hardware Summit 2023 is truly humbling and reflects my commitment to open-source hardware. Getting recognized as one of the inspiring women in STEM by Girls STEMpede in 2022 is a cherished moment, highlighting my dedication to supporting women in science and technology.

I'm grateful to be featured as a princess in the "Princesses with Powertools" calendar for 2023, aiming to inspire young girls and letting them know that they don't have to choose between being an engineer or a princess.

Additionally, being featured as a Maker on the Hackster Blog has allowed me to share my enthusiasm for electronics and open-source projects with a wider audience. But the biggest accomplishment is always after I teach a workshop or someone who's never done anything with electronics sees my project and makes their first project inspired by my work.

On another note, it's always a proud moment when designs I've worked on for Ryobi are used every day by people everywhere. Watching trials in strawberry farms, where the robots I helped design pick strawberries flawlessly at the crack of dawn, has been another source of pride. Integrating the very first custom PCB design, into the very first all-electric RV designed for commercial use, is also a significant professional milestone.

What was more challenging than you anticipated?

What proved more challenging than initially anticipated was the realization that many essential practical skills in Electrical Engineering were not covered in formal education. Tasks like testing, debugging, PCB design, and soldering were areas I had to independently learn with limited accessible resources. The software necessary for designing PCBs often comes with a cost, so as a student, I had to rely on free trials and teach myself during that trial period to acquire the skills required for job applications. Despite these hurdles, this experience fueled my determination to make PCB design more accessible. It inspired a commitment to open-source designs and an eagerness to teach these skills in a way that breaks down the barriers to entry. In addition to these challenges, I observed a pervasive myth of the "math brain" and gender stereotypes perpetuating the notion that women are not naturally inclined to excel in hardware concepts, serving as another barrier that I actively work to challenge. This misconception discourages individuals, particularly women, from pursuing these fields, reinforcing harmful biases.

Furthermore, I noticed a pattern among engineers who claimed to have a natural knack for technology – many had close family members who were engineers and had been exposed to these concepts early on. This observation highlights an additional layer of inequality, and emphasizes the importance of dismantling such biases to foster a more inclusive environment in Electrical Engineering.

What do you wish you knew before you chose this career?

Before delving into Electrical Engineering, I wish I had been more aware of the myriad and impactful applications within this field. Throughout my studies, the depth of potential to create solutions that contribute to a cleaner and healthier planet, as well as the artistic possibilities that engineering holds, eluded me. Discovering these aspects later in my journey significantly altered my perspective on engineering, enriching my appreciation for its inherent creativity.

However, one aspect that took me by surprise was the gender bias embedded within the field, and the accompanying imposter syndrome. In an environment that thrives on experimentation, learning from failures, and fostering a growth mindset, imposter syndrome can prove particularly detrimental. It erects a barrier, impeding exploration, and hindering the pursuit of developing skills further. Recognizing and addressing these challenges early on would have better equipped me for navigating my career path.

Additional ways you promote women in engineering and/or STEM.

I actively promote women in engineering and STEM through various initiatives. One significant avenue is creating open-source electronic projects that fuse art with electronics, such as PCB jewelry and light-up dresses. These designs are shared in open-source formats, encouraging others, especially girls and women, to build and learn from them. Additionally, I conduct introductory electronics workshops tailored for girls and have established Sparkletronics, a small business dedicated to developing electrical engineering educational content, and beginner-friendly kits. These kits, shaped in fun forms like flowers and butterflies, aim to eliminate the intimidation often associated with learning electronics, thereby making the subject more accessible.

As part of my advocacy efforts, I participate actively in open hardware conferences, where I showcase my work, deliver talks, and champion diversity and inclusion in these spaces. This involvement serves as a platform to amplify the representation of women in engineering and STEM fields, while fostering an inclusive and supportive community.

Anything else you would like to share with the reader?

Don't be afraid to change your mind or your path; explore and experiment with your career. Follow your heart and always bet on yourself, even when no one else is, and especially when someone is actively telling you that you can't do something; because here I am, telling you that you absolutely can, and you absolutely must!

Connect with Ayesha:

@AYESHA.IFTIQHAR

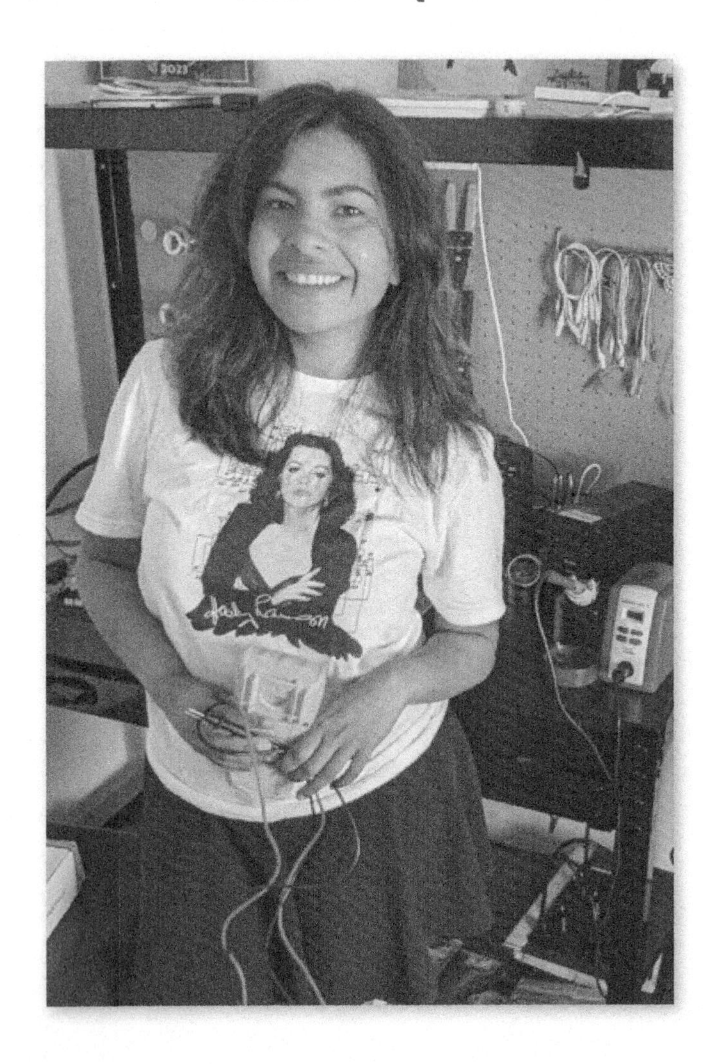

About the Author

Dr. J. A. Sanchez, better known as Justina Sanchez, is a San Diego native, a wife, mom, best-selling author, and Quality Engineer III. She has been in the Tech Industry for 19 years and is a big supporter of advancing women in engineering.

From testing and evaluating products before they go to market, to inspecting manufacturing facilities and auditing the labs that support these evaluations. Justina has a diverse background in engineering, and she continues to grow in her technical expertise while sharing her knowledge and experiences with others.

Justina has found that some of her greatest lessons have come through mentoring others and also being mentored by those further ahead. She loves others with the love of Jesus, and is the biggest cheerleader to those running alongside her. Justina enjoys making memories with family and friends, as well as traveling and experiencing new cultures. She also has a passion for learning, building relationships, and providing opportunities for others.

Connect with Extraordinary Engineers

Connect with Justina

Visit us online at www.extraordinaryengineers.com

Made in the USA
Coppell, TX
06 May 2025